石油石化企业现场安全督导系列丛书

基本 HSE 风险因素及管理

Fundamental HSE Risk Factors and Management

赵宏展　贺晓珍　杨意峰　编著

中国石化出版社

内 容 提 要

本书为《石油石化企业现场安全督导系列丛书》之一,介绍了 HSE 风险管理工具和过程、HSE 风险贡献因素和升级因素、旅程管理和道路交通安全、人员健康和保护共计四组知识模块,包含工作危害分析、作业许可、工作场所 HSE 会议等 15 个专题,可作为做好管理人员现场安全督导需要了解的支持性信息。

本书内容为中英文对照,便于读者将中文和英文结合在一起阅读,以便准确理解其内容。

本书可用于石油石化企业各级管理人员自学或培训,也可作为其他行业的管理人员以及石油石化大专院校工科专业及其他院校安全相关专业师生的安全管理类参考书。

图书在版编目(CIP)数据

基本 HSE 风险因素及管理 / 赵宏展,贺晓珍,杨意峰编著.—北京:中国石化出版社,2019.8
(石油石化企业现场安全督导系列丛书)
ISBN 978-7-5114-5450-8

Ⅰ.①基… Ⅱ.①赵… ②贺… ③杨… Ⅲ.①石油化工企业—安全管理 Ⅳ.①TE687.2

中国版本图书馆 CIP 数据核字(2019)第 169598 号

未经本社书面授权,本书任何部分不得被复制、抄袭,或者以任何形式或任何方式传播。版权所有,侵权必究。

中国石化出版社出版发行
地址:北京市东城区安定门外大街 58 号
邮编:100011 电话:(010)57512500
发行部电话:(010)57512575
http://www.sinopec-press.com
E-mail:press@sinopec.com
北京柏力行彩印有限公司印刷
全国各地新华书店经销

*

850×1168 毫米 32 开本 6.375 印张 141 千字
2019 年 11 月第 1 版 2019 年 11 月第 1 次印刷
定价:36.00 元

前言
Preface

　　企业健康安全环境(HSE)管理工作的属地管理和直线责任原则已经深入人心,很多企业也高度重视"有感领导"(指有安全认知的领导),不断强化各级管理人员对HSE工作的示范作用。越来越多企业的管理人员经常深入生产作业现场,实地调研和指导HSE管理工作,践行"有感领导"。随着企业HSE管理体系越来越成熟,不同角度、不同深度、不同方式的各级管理人员开展的HSE检查和HSE审核也越来越丰富多样,对于生产作业场所的HSE风险管控起到了至关重要的作用。

　　上述背景产生了两个问题:一是石油和石化企业各级管理人员在企业HSE管理体系框架下如何更精准有效地发力?二是石油和石化企业有哪些共性的HSE知识模块值得总结和提炼?

　　编著者基于10余年的海外项目工作经验,在广泛调研国内外大量文献资料的基础上,通过跟诸多资深项目管理人员和资深HSE专业人员的沟通交流,编写了本丛书。本丛书呈上了对上述两个问题的初步回应,推荐了基于"安全交谈(Safety Conversation)"的"管理人员现场安全督导方案(Leadership Site Visit Program)",包括:具体做法、注意事项、支持性信息(知识模块)。

前言

在安全氛围(Safety Climate)的三"心"模型❶基础上,我们坚信:基于"安全交谈"的管理人员现场安全督导,可以通过管理人员对于安全问题的"关心",激励员工对于安全问题的"上心",并最终营造企业或项目对于安全问题的"上下一心"。

人们期待"安全保障",并提出了"安全保障"的需求和要求。实际上,只有"共建安全"才能获得更加牢靠的"安全保障",因为"安全保障"的需求者和供给者是不可分离的。在属地管理、直线责任、"有感领导"的基础上,人们还必须坚持"安全共建"和"安全共享"的原则,即每个人都坚信"我需要安全,安全需要我"。

本丛书呈上的中英文对照的支持性信息(知识模块)共分为八组,分别是:

(1) HSE 风险管理工具和过程;

(2) HSE 风险贡献因素和升级因素;

(3) 旅程管理和道路交通安全;

(4) 人员健康和保护;

(5) 工作场所常见危害;

(6) 潜在高后果活动;

(7) 危险有害物质;

(8) 火灾爆炸保护。

---

❶于广涛,李永娟. 安全氛围三"心"模型的构建与检验[J]. 中国安全科学学报, 2009, 19(9): 28.

前言

本丛书"臻选"的支持性信息(知识模块)融入了以下内容：(危害)控制层序(Hierarchy of Controls)的原则、设计保障安全(Prevention through Design)的做法、领结图(Bowtie Diagram)的精髓、保护层分析法(Layer of Protection Analysis)的思路、屏障(Barrier)的要义。此外，安全例证(Safety Case)不仅仅是英国等发达国家法律规定的具体做法，更是一种风险管理思路。危险轨迹(Line of Fire)既是生产作业场所的一种"常态"，也是很多人身伤害的本质，目前已经作为国际油气生产商协会(IOGP)在2018年发布的九条保命法则(Life-Saving Rules)之一而更加受到重视。管理巡查(Management Walkthrough)、管理层检查(Management Inspection)和管理审核(Management Audit)，三者密切相关，也经常被混淆，本丛书对其区别和联系也做出了解读。

管理人员现场安全督导推荐做法、常用安全模型以及八组支持性信息(知识模块)等内容顺序编排在本丛书的相应分册当中。四个分册的名称具体如下：

- 《安全督导推荐做法及常用安全模型》
- 《基本HSE风险因素及管理》
- 《工作场所常见危害及潜在高后果活动》
- 《危险有害物质及火灾爆炸保护》

本丛书的英文内容参考了以下政府组织和行业协会公开出版的英文文献资料，在此表示诚挚感谢！

前言

◇ 英国健康安全执行局(UK HSE, UK Health and Safety Executive)

◇ (美国)职业安全与卫生管理局(OSHA, Occupational Safety and Health Administration)

◇ 美国化学工程师协会(AIChE, American Institute of Chemical Engineers)

◇ 国际油气生产商协会(IOGP, International Association of Oil & Gas Producers)

◇ 国际钻井承包商协会(IADC, International Association of Drilling Contractors)

本丛书的写作还参考了其他大量国内外文献资料,在此对原著者也深表感谢!

回望过去,本丛书英文内容的最终定稿是一个漫长的积淀和筛选过程,历时10余年时间。2010年1月,编著者之一刚刚取得安全工程博士学位,作为唯一的中方人员,有幸被指派组建一支国际化的HSE团队,服务于一个员工来自30多个国家的中东地区油田开发生产项目。先后有来自英国、美国、法国、加拿大、澳大利亚、新西兰等国家的60多位资深HSE专业人士以油田员工或知名咨询公司顾问的身份,成为这支国际化HSE团队的一员。2012年12月和2013年6月,本丛书的另两位编著者先后加入这支HSE团队,成为协调和领导这支国际化HSE团队的核心力量,但英文仍然是覆盖整个HSE团

前　言

队的唯一工作语言。在开放、包容、交流、共享、相互尊重的基础上，这支 HSE 团队整体所能提供的 HSE 支持和服务实现了最大程度的国际接轨。总之，本丛书可以说是编著者经过 10 余年开放融通的海外项目磨砺后交上的一份答卷。

本丛书"臻选"的支持性信息(知识模块)采用中英文对照的形式编写和排版，目的是希望读者能够将中文和英文结合在一起阅读，提取其中的"最大公约数"，从而更加精准地理解本丛书所提供信息和知识的本意，更好地辅助和支持管理人员开展现场安全督导。

本丛书最终得以公开出版得到了众多资深管理人员、中外专家和 HSE 同仁的支持和参与，凝聚了众人的智慧和心血，编著者不过是在知识的沙滩上捡拾贝壳的三人，我们虔诚地把认为有价值的知识贝壳收集起来，经过筛选和加工以中英文双语的形式呈现给广大读者。感谢默默无闻支持编著者完成本丛书的所有人！

本丛书主要适用于石油石化企业各级管理人员自学或培训之用，也可作为其他行业的管理人员以及石油石化院校工科专业及其他院校安全相关专业师生的安全管理类参考书。

管理人员现场安全督导工作有着很强的知识性和实践性，书中难免存在疏漏之处，敬请批评指正。如有任何意见和建议，欢迎联系以下电子邮件地址：safetyview@163.com。

目录

1 HSE 风险管理工具和过程

1.1 工作危害分析 /3
1.1.1 督导提示 /3
1.1.2 知识准备 /5
1.1.3 参考资料 /9

1.2 作业许可 /11
1.2.1 督导提示 /11
1.2.2 知识准备 /13
1.2.3 相关保命法则 /23
1.2.4 参考资料 /23

1.3 工作场所 HSE 会议 /25
1.3.1 督导提示 /25
1.3.2 知识准备 /27

1.4 应急管理和危机管理 /37
1.4.1 督导提示 /37
1.4.2 知识准备 /41

目录

2 HSE风险贡献因素和升级因素

2.1 健康适岗 /55
2.1.1 督导提示 /55
2.1.2 知识准备 /57
2.1.3 参考资料 /63

2.2 恶劣天气 /65
2.2.1 督导提示 /65
2.2.2 知识准备 /67

2.3 夜间作业 /71
2.3.1 督导提示 /71
2.3.2 知识准备 /73
2.3.3 参考资料 /79

2.4 独自作业 /81
2.4.1 督导提示 /81
2.4.2 知识准备 /83
2.4.3 参考资料 /89

2.5 变更管理 /91
2.5.1 督导提示 /91
2.5.2 知识准备 /93
2.5.3 参考资料 /103

3 旅程管理和道路交通安全

3.1 旅程管理 /107
3.1.1 督导提示 /107

3.1.2 知识准备 /109
3.1.3 相关保命法则 /117
3.1.4 参考资料 /117

3.2 道路交通安全 /119
3.2.1 督导提示 /119
3.2.2 知识准备 /123
3.2.3 相关保命法则 /131
3.2.4 参考资料 /131

4 人员健康和保护

4.1 食品安全 /135
4.1.1 督导提示 /135
4.1.2 知识准备 /137

4.2 热应激 /149
4.2.1 督导提示 /149
4.2.2 知识准备 /151
4.2.3 参考资料 /157

4.3 噪声与听力保护 /157
4.3.1 督导提示 /157
4.3.2 知识准备 /159
4.3.3 参考资料 /167

4.4 个体防护装备 /167
4.4.1 督导提示 /167
4.4.2 知识准备 /171
4.4.3 参考资料 /175

目录

附录

附录1　管理人员现场安全督导提示卡样例　　/179

附录2　缩略语　　/182

附录3　中英文对照词汇表　　/184

Contents

1 HSE RISK MANAGEMENT TOOLS AND PROCESS

1.1 Job Hazard Analysis /2
   1.1.1 Prompts /2
   1.1.2 Reading /4
   1.1.3 References /8

1.2 Permit to Work /10
   1.2.1 Prompts /10
   1.2.2 Reading /12
   1.2.3 Related Life-Saving Rules /22
   1.2.4 References /22

1.3 Workplace HSE Meetings /24
   1.3.1 Prompts /24
   1.3.2 Reading /26

1.4 Emergency Management and Crisis Management /36
   1.4.1 Prompts /36
   1.4.2 Reading /40

2 CONTRIBUTION FACTORS AND ESCALATION FACTORS FOR HSE RISK

2.1 Fitness to Work /54
2.1.1 Prompts /54
2.1.2 Reading /56
2.1.3 References /62

2.2 Adverse Weather /64
2.2.1 Prompts /64
2.2.2 Reading /66

2.3 Night Working /70
2.3.1 Prompts /70
2.3.2 Reading /72
2.3.3 References /78

2.4 Work Alone /80
2.4.1 Prompts /80
2.4.2 Reading /82
2.4.3 References /88

2.5 Management of Change /90
2.5.1 Prompts /90
2.5.2 Reading /92
2.5.3 References /102

3 JOURNEY MANAGEMENT AND LAND TRANSPORT SAFETY

3.1 Journey Management /106
3.1.1 Prompts /106

3.1.2	Reading	/108
3.1.3	Related Life-Saving Rules	/116
3.1.4	References	/116

3.2 Land Transport Safety /118

3.2.1	Prompts	/118
3.2.2	Reading	/122
3.2.3	Related Life-Saving Rules	/130
3.2.4	References	/130

4 PERSONAL HEALTH AND PROTECTION

4.1 Food Safety /134

4.1.1	Prompts	/134
4.1.2	Reading	/136

4.2 Heat Stress /148

4.2.1	Prompts	/148
4.2.2	Reading	/150
4.2.3	References	/156

4.3 Noise and Hearing Conservation /156

4.3.1	Prompts	/156
4.3.2	Reading	/158
4.3.3	References	/166

4.4 Personal Protective Equipment /166

4.4.1	Prompts	/166
4.4.2	Reading	/170
4.4.3	References	/174

APPENDIX

Appendix 1	**Leadership Site Visit Prompt Card Sample**	/178
Appendix 2	**Abbreviations and Acronyms**	/182
Appendix 3	**English-Chinese Vocabulary**	/184

HSE RISK MANAGEMENT TOOLS AND PROCESS

HSE风险管理工具和过程

1.1 Job Hazard Analysis

1.1.1 Prompts

(1) Have all routine job hazards been identified, risks assessed and used to establish safe methods of work?

(2) Are all non-routine job hazards identified, risk assessed and used to establish safe methods of work as part of the PTW system?

(3) Is a JHA procedure covering routine and non-routine work activities in place and operational?

(4) Are JHAs being performed using the standard JHA form from the HSE-MS?

(5) Are personnel aware of how to use a best practice JHA for routine jobs?

(6) Are best practice JHAs available and being actively used, reviewed and updated for routine jobs?

(7) Are assessments of the daily hazards (such as weather conditions, other work being completed in the vicinity, etc.) being performed for routine jobs that use best practice JHA?

(8) Are the JHA forms being used correctly, are hazards and risks being recorded correctly and adequate control measures put into place?

(9) Is the work team actively involved in the discussion of hazards and hazard controls during preparation of the JHA?

1.1 工作危害分析

1.1.1 督导提示

(1) 是否已识别出所有常规作业❶危害？是否对危害进行了风险评估？是否据此确立了安全的工作方法？

(2) 作为作业许可制度(PTW)的组成部分，是否识别出了所有非常规作业危害？是否对危害进行了风险评估？是否据此确立了安全的工作方法？

(3) 涵盖常规和非常规作业活动的工作危害分析(JHA)程序是否到位并有效实施？

(4) 正在执行的工作危害分析(JHA)是否使用了健康安全环境管理体系(HSE-MS)文件规定的标准表格？

(5) 针对常规作业，作业人员是否清楚如何使用工作危害分析(JHA)范例？

(6) 针对常规作业，是否有现成的工作危害分析(JHA)范例？工作危害分析(JHA)范例是否被主动使用、审查并更新？

(7) 针对使用工作危害分析(JHA)范例的常规作业，是否对日常(动态)危害(例如天气状况、在附近区域正在进行的其他作业活动等)进行适时评估？

(8) 是否正在正确地使用工作危害分析(JHA)表格？是否正确地记录危害及其风险？是否制定了合乎需要的控制措施？

(9) 在准备工作危害分析(JHA)时，作业班组人员是否积极参与危害(识别)和危害控制措施的讨论？

---

❶根据上下文，Job 翻译为工作或者是作业。

(10) Is the temporary override of HSE Critical Equipment and Systems (HSECES) identified within the JHA, including a maximum allowed time for the override? Does the JHA state what happens after the allowed time is exceeded? Does the JHA refer to the override log?

(11) Are best practice JHAs available and being actively used by the work team?

(12) Is the process for archiving JHAs issued under the PTW in place and operational?

(13) Are JHAs archived under the PTW easily retrievable for similar new jobs?

(14) Have site personnel received training on the requirements for JHA?

(15) Is a process in place to ensure that personnel who work with JHA are competent in its use and application?

1.1.2 Reading

A Job Hazard Analysis (JHA) is a technique that focuses on the steps of a job as a way to identify hazards before they occur. It focuses on the relationship between the worker, the job, the tools, and the work environment.

JHA breaks down a job into a number of steps and identifies the hazards from each step. Measures can then be put in place to control each of the hazards identified. The amount of effort required to perform a JHA is commensurate to the complexity and level of risk of the work to be performed.

(10) 工作危害分析(JHA)是否识别了 HSE 关键设备和系统(HSECES)的临时旁路？是否包括最大允许的旁路时间？工作危害分析(JHA)是否阐明了超过允许的(旁路)时间后会发生什么？工作危害分析(JHA)是否参考了旁路日志？

(11) 工作危害分析(JHA)范例是否可获得并被作业班组人员积极使用？

(12) 是否建立了合适的程序对作业许可(PTW)包含的工作危害分析(JHA)进行归档？该程序是否有效运行？

(13) 针对相似的新作业，已归档的作业许可(PTW)包含的工作危害分析(JHA)是否易于取用？

(14) 现场人员是否接受过工作危害分析(JHA)要求的培训？

(15) 是否建立了程序确保涉及工作危害分析(JHA)的人员有能力使用和应用工作危害分析(JHA)？

1.1.2 知识准备

工作危害分析(JHA)是在工作步骤展开前，按工作任务步骤，逐步识别危害的方法。它关注工作人员、工作内容、工具以及作业环境之间的相互关系。

工作危害分析(JHA)将工作任务分成几个步骤，识别出每一个步骤可能的危害，然后针对每一项危害采取有效控制措施。开展工作危害分析(JHA)所需的工作量与要执行工作任务的复杂性和风险级别相对应。

JHA, when associated with non-routine jobs in hazardous locations, becomes a formal part of the Permit to Work (PTW) system. JHA when associated with routine jobs becomes part of best practice JHAs which can be used as the basis for developing activity specific Standard Operating Procedures (SOP) or Work Instructions (WI).

Supervisors can use the findings of a JHA to eliminate and prevent hazards in their workplaces. JHA has many other benefits including fewer worker injuries and illnesses; safer, more effective work methods; reduced workers' compensation costs and increased worker productivity.

For JHA to be effective, management must demonstrate its commitment to health and safety and follow through to correct any uncontrolled hazards identified.

- **A JHA procedure should be in place and operational**

JHA is a mandatory requirement of the PTW system and is required for all non-routine jobs performed in hazardous conditions or locations.

JHA for routine jobs utilizes best practice JHA. A best practice JHA is one that reflects the best practice hazard management that can be used as a reference for performing a routine job.

A best practice JHA, should be prepared by site supervisors and workers usually at the first instance of performing the routine job. The development of best practice JHAs may also be facilitated by HSE. Best practice JHAs should be periodically checked by HSE to ensure the JHA meets technical and quality requirements. The HSE approved best practice JHA should then be used as a reference during Toolbox Talks (TBT) for all workers who subsequently perform the routine job.

在危险位置/地点/场地执行非常规作业时，工作危害分析(JHA)是作业许可(PTW)制度不可或缺的一部分。与常规工作相关的工作危害分析(JHA)，是工作危害分析(JHA)范例的一部分。工作危害分析(JHA)范例可用作开发作业活动具体的标准作业程序(SOP)或作业指导书(WI)的基础。

工作危害分析(JHA)的结果可以帮助现场主管人员消除和防控工作场所的危害。工作危害分析(JHA)还有许多好处，包括：更少的工伤和疾病；更安全、更有效的工作方法；降低工作人员的赔偿费用并提高其生产率。

为了有效实施工作危害分析(JHA)，管理层必须展示其对健康和安全的承诺，并把纠正识别出的任何危害的不受控进行到底。

- **应建立工作危害分析(JHA)程序并有效实施**

工作危害分析(JHA)是作业许可(PTW)制度的强制要求，在危险环境或危险位置/地点/场地开展非常规作业时必须执行工作危害分析(JHA)。

常规作业使用工作危害分析(JHA)范例。工作危害分析(JHA)范例是用来作为参考的工作危害分析(JHA)，它展示的是执行常规作业时危害管理的最佳做法。

通常情况下，工作危害分析(JHA)范例是由现场主管人员和作业人员在第一次开展某项常规作业时准备的。开发工作危害分析(JHA)范例也可由HSE人员进行协助。HSE人员应定期审查工作危害分析(JHA)范例以确保其符合技术和质量要求。由HSE人员批准的工作危害分析(JHA)范例可作为工具箱会议(TBT)的参考资料，供随后执行此项常规作业的人员参考使用。

Suggestions for improvement can be made for the relevant best practice JHA during each TBT (Toolbox Talk).

Site conditions can change from day to day. Methods for performing the job can improve through new technology or knowledge. Best practice JHAs should be re-evaluated and updated periodically to reflect improvements.

- **A process should be in place to ensure that personnel involved in JHA and work execution are competent**

Supervisors should be competent to lead and prepare JHA as part of PTW system.

JHA should identify general and job specific training needs associated with the management of work related hazards and their controls.

Induction and basic level JHA training should be provided as a minimum to all workers with JHA and job performing roles and responsibilities within the PTW system. Workers should have high level of hazard awareness, understand the requirements of JHA and understand how it improves the way they work.

1.1.3 References

OSHA, Job Hazard Analysis, OSHA 3071, 2002(Revised).

在每次工具箱会议(TBT)期间,所有人员都可针对相关工作危害分析(JHA)范例提出改进建议。

由于现场条件可能每天都在变化,执行作业的方法随着新技术或新知识在不断改进,需要定期对工作危害分析(JHA)范例进行重新评估和更新以反映相关改进。

- **应建立适当程序确保参与工作危害分析(JHA)和执行作业的人员胜任**

现场主管人员应具备遵循作业许可(PTW)制度以及牵头并准备工作危害分析(JHA)的能力。

工作危害分析(JHA)应确定出跟工作危害及其防控措施管理相关的通用和特定培训需求。

作为最低要求,应该为所有在作业许可(PTW)制度中承担工作危害分析(JHA)以及任务执行角色和责任的人员,提供入门以及基本的工作危害分析(JHA)培训。工作人员应该具有高度的危险意识,清楚工作危害分析(JHA)的要求并清楚工作危害分析(JHA)如何改善自己的工作方式。

1.1.3 参考资料

OSHA, Job Hazard Analysis, OSHA 3071, 2002(Revised).

1.2 Permit to Work

1.2.1 Prompts

(1) Is a PTW system to control non-routine high risk potential work activities in place and operational?

(2) Is a process to control work areas and area authorisation of permits in place and operational?

(3) Is a process to control permit validity, suspension and extension in place and operational?

(4) Is a process to control the interactions between multiple live permits in place and operational?

(5) Are all the live permits displayed at the Permit Control Facility?

(6) Is the process to communicate the permit process, risk assessment and control measures to workers and other stakeholders in place and operational?

(7) Is a process to control task risk assessments through Job Hazard Analysis(JHA)in place and operational?

(8) Are permit copies available at the point of use?

(9) For live permits, have all personnel involved in the job been instructed on the permit requirements and do they understand their role and responsibilities?

(10) Is a process to control and permit mechanical and energy isolation in place and operational?

1.2 作业许可

1.2.1 督导提示

(1) 是否建立并实施了控制非常规潜在高风险作业活动的作业许可(PTW)制度?

(2) 是否建立并实施了控制作业区域及(作业)许可区域核准的程序?

(3) 是否建立并实施了控制(作业)许可有效性、暂停及延期的程序?

(4) 是否建立并实施了控制多个有效(作业)许可之间相互影响的程序?

(5) 是否将所有有效(作业)许可张贴在"(作业)许可控制设施"上?

(6) 是否建立并实施了跟作业人员及其他相关方沟通(作业)许可流程、风险评估和控制措施的程序?

(7) 是否建立并实施了通过工作危害分析(JHA)进行任务风险评估的程序?

(8) 是否可以在使用(作业许可的)作业点获得(作业)许可副本?

(9) 对于有效(作业)许可,是否将许可要求传达给了参与作业的所有人员?参与作业的所有人员是否都清楚自己的角色与职责?

(10) 是否建立并实施了控制和许可机械与能量隔离的程序?

(11) Is a process to control and permit the temporary override of Safety Critical Equipment(System) in place and operational?

(12) Is a process to control, permit and certify gas testing in place and operational?

(13) Is a process to control and permit confined space entry in place and operational?

(14) Is a process to control and permit lifting operations in place and operational?

(15) Is a process to control and permit excavation work in place and operational?

(16) Is a process in place to ensure that safety critical personnel who work with the PTW system are competent in its use and application?

(17) Have personnel on site received training and certification for their role in the PTW system?

(18) Is a process to monitor, audit and improve the performance of the PTW system in place and operational?

1.2.2 Reading

A Permit to Work(PTW) system is an integral part of a HSE management system (HSE-MS) and is a formal recorded process used to control work which is identified as potentially hazardous. It is also a means of communication among site management, supervisors and operators and those who carry out the hazardous work.

（11）是否建立并实施了控制和许可临时旁路(停用)安全关键设备(系统)的程序？

（12）是否建立并实施了控制、许可和(书面)证明气体检测的程序？

（13）是否建立并实施了控制和许可进入受限空间的程序？

（14）是否建立并实施了控制和许可吊装作业的程序？

（15）是否建立并实施了控制和许可动土作业的程序？

（16）是否建立了程序确保执行作业许可(PTW)制度的安全关键人员在使用和应用作业许可(PTW)制度方面是胜任的？

（17）现场人员是否接受了在作业许可(PTW)制度中所承担角色的培训并合格？

（18）是否建立并实施了监测、审核和改进作业许可(PTW)制度实施情况的程序？

1.2.2 知识准备

作业许可(PTW)制度是 HSE 管理体系(HSE-MS)的有机组成部分。它是正式的(确认危害及其控制有效性并)加以记录的过程，用于控制被确定为具有潜在危险性的作业。它也是一种现场负责人、主管人员、操作人员以及那些从事危险作业人员之间的沟通方式。

A PTW system aims to ensure that proper consideration is given to the risks of a particular job or simultaneous activities on site. Whether it is manually or electronically generated, the permit is a detailed document that authorizes certain people to carry out specific work at a specific site at a certain time, and that sets out the main precautions needed to complete the job safely.

Job Hazard Analysis (JHA) is an essential requirement of a PTW system.

A PTW is not simply a permission to carry out a dangerous job. It is an essential part of a system which determines how that job can be carried out safely, and helps communicate this to those doing the job. It should not be regarded as an easy way to eliminate hazard or reduce risk. The issue of a permit does not, by itself, make a job safe, that can only be achieved by those preparing for the work, those supervising the work and those carrying it out in line with the PTW requirements.

The PTW system includes certificates on an as needed basis. Certificates include work at height, lifting, excavation, isolation, confined space entry and radiography. Certificates are issued based on the JHA conducted before any work is undertaken. The PTW system should ensure that authorised and competent people have thought about foreseeable risks and that such risks are avoided by using suitable precautions.

PTW systems are normally required for controlling complex non-routine (e.g. inspection, maintenance, repair, etc.) jobs in hazardous areas.

作业许可(PTW)制度旨在确保现场特定作业或同时进行的活动的风险被恰当考虑。无论是手写还是电子生成，(作业)许可都是一份详细的文件，授权已确定的人员在已确定的时间和明确的地点执行明确的作业，并列出安全作业所需的主要防范措施。

对作业许可(PTW)制度来讲，工作危害分析(JHA)是不可或缺的。

一个作业许可(PTW)不仅仅是批准一项危险作业的执行，它是决定如何安全执行这项(危险)作业的一个系统的必要组成部分，有助于将上述系统的情况传达给作业人员。不应把作业许可(PTW)本身当成是消除危害或降低风险的简便方法。签发作业许可本身并不能保证作业安全，作业安全需要由作业准备人员、作业监督人员和作业执行人员都严格遵守作业许可(PTW)的要求来实现。

根据实际需要，作业许可(PTW)制度包含不同的专项作业单，具体包括高处作业单、吊装作业单、动土作业单、隔离作业单、进入受限空间作业单和射线作业单等。专项作业单的签发需基于作业开始之前的工作危害分析(JHA)。作业许可(PTW)制度应确保被授权的胜任人员已经思考过可预见的风险并通过采取适当的防范措施规避这些风险。

作业许可(PTW)制度通常用于控制在危险区域(执行的)的复杂的非常规作业(例如：检查、维护、修理等)。

1 HSE RISK MANAGEMENT TOOLS AND PROCESS

- **A PTW system should be in place and operational**

In order to manage potentially hazardous work, the site should operate a formal PTW system which meets the minimum requirements of the HSE-MS and relevant specifications. The PTW system should be recognised throughout the site as being mandatory for hazardous non-routine work activities.

There should be clear identification of who may authorise particular jobs (and any limits to their authority) and who is responsible for specifying the necessary precautions.

There should be a clear and standardised identification of jobs, risk assessments, permitted job duration and supplemental or simultaneous activity and control measures.

Particular attention to the PTW system should be made during combined or simultaneous operations (SIMOPS) to ensure that work undertaken does not compromise safety. In particular, the following are examples of types of job where supplementary certificates should be considered:

① Hot work which may generate heat, sparks or other sources of ignition.

② Work which may involve disrupting containment of a flammable, toxic or other dangerous substance and/or pressure system.

③ Work on high voltage electrical equipment or other work on electrical equipment which may give rise to danger.

④ Entry and work within tanks and other confined spaces.

⑤ Work involving the use of hazardous/dangerous substances, including radioactive materials and explosives.

- **作业许可(PTW)制度应到位并有效运行**

为了管理具有潜在危险性的作业,作业现场应运行正式的作业许可(PTW)制度,该制度应符合 HSE 管理体系(HSE-MS)和相关规范的最低要求。作为对危险的非常规作业活动的强制性要求,作业许可(PTW)制度应在作业现场全面实施。

应明确规定由谁授权什么特定作业(包括对其权限的任何限制)以及谁负责确定必要的防范措施。

对作业(内容)、风险评估、批准的作业时间、追加或同时进行的作业活动以及控制措施的确定,应清晰且标准化。

应特别注意针对联合或同步作业❶的作业许可(PTW)制度,确保所进行的作业不会危及安全。具体来讲,以下列举了应考虑专项作业单的作业类型:

① 可能产生热量、火花或其他点火源的热作业。
② 可能涉及易燃、有毒、其他危险物质和/或压力系统防护措施破坏的作业。
③ 高压电气设备作业或其他可能导致危险的电气设备作业。
④ 进入大罐和其他受限空间内及在其中作业。
⑤ 使用危险有害物质(包括放射性材料和火工品)的作业。

---

❶ SIMOPS 也称作"同时作业"或"交叉作业"。

⑥ Pressure testing as part of process isolation.

⑦ Work involving temporary overrides of safety critical equipment (system).

⑧ Work at height.

⑨ Any other potentially high-risk operation.

The PTW system should be tested through "sampling" the permit development, issue, authorisation and permitted work. It should be confirmed that the level of detail within the supporting forms, JHA and other supporting documents (if any) is commensurate to the level of risk posed by the work. Permits (hot work and cold work) and the supplementary certificates (isolation, confined space entry, lifting, work at height, excavation, etc.) that are part of the PTW system should be reviewed for adequacy and completeness.

- **A process should be in place to ensure that personnel who work in the PTW system are competent**

An assessment of competency should cover practical skills and reasoning as well as knowledge. Training should focus on use of the PTW system, but must also ensure that the individual understands the working environment, the hazards associated with it, and more importantly, the controls required to appropriately manage the risks presented by those hazards.

Effective training is essential to achieve quality and consistency in the use of the PTW system. There should be successive levels of training (i.e. ranging from basic to advanced) for those involved. Training provides the foundation for effective implementation of a PTW system and supports user competence development.

⑥ 作为工艺隔离一部分的压力测试。
⑦ 临时停用安全关键设备(系统)的作业。
⑧ 高处作业。
⑨ 任何其他具有潜在高风险的作业。

应通过对作业许可(PTW)准备、签发、授权和执行等环节的抽样对作业许可(PTW)系统进行检查。应确认支持性表格、工作危害分析(JHA)以及其他支持性文件(如果有的话)的详细程度与作业风险等级是否匹配。应审查作为作业许可(PTW)系统组成部分的作业许可(热作业和冷作业)和专项作业单(隔离、进入受限空间、吊装、高处、动土等)的充分性和完整性。

- **应建立程序确保作业许可(PTW)制度涉及的人员胜任**

能力评估不仅应涵盖知识,还应涵盖实践能力和思维能力。培训应侧重于作业许可(PTW)制度的实施,但还必须确保每一位作业人员理解工作环境以及与工作环境相关的危害。更重要的是,要理解为恰当管理那些危害呈现的风险所需要的措施。

有效的培训对于作业许可(PTW)制度的实施质量和实施一致性至关重要。对于所有相关人员,应该进行不同级别的接续培训(如从基础到高级)。培训为有效实施作业许可(PTW)制度提供了基础,并为实施人员的能力发展提供支持。

Training is the first step for all PTW users. A PTW has a range of users with varying roles and responsibilities, ranging from Permit Authorities who authorise permits down to technicians who participate in the Toolbox Talks and execute parts of the work. Training is required for all personnel with PTW roles and responsibilities in order to improve understanding of the system, encourage ownership and support participation.

Training should cover:

① The principles of a PTW system.

② When permits are required.

③ An understanding of the types of permits, supporting certificates and other documentation (e.g. JHA and method statements).

④ Responsibilities and competence requirements for authorised personnel.

⑤ Responsibilities of permit users.

⑥ Lessons from incidents associated with PTW.

⑦ Findings from PTW System audit and review.

Refresher training should normally be required at specific intervals (one year/two years), after a change in the system or following the recommendations of a system audit.

- **A process should be in place to monitor and audit the PTW system**

In addition to checks carried out by issuers, PTW monitoring checks should be undertaken by site management and supervisors to validate compliance with detailed work instructions and control measures. Information gained from permit monitoring should be used to reinforce safe working practices on site.

培训是所有作业许可(PTW)使用人员的第一步。一个作业许可(PTW)涉及一系列具有不同角色和职责的使用人员，从作业许可的批准人员到参加工具箱会议并执行部分作业的技术人员。在作业许可(PTW)制度中扮演角色和行使职责的人员都需要进行培训，以提高(他们)对作业许可(PTW)制度的理解、激发其责任感、强化其参与度。

培训应涵盖：

① 作业许可(PTW)制度的原则。
② (作业)许可的适用性。
③ 理解(作业)许可的类型、支持作业许可的专项作业单和其他文件[例如：工作危害分析(JHA)和作业方法说明]。
④ 对授权人的职责和能力要求。
⑤ (作业)许可使用人员的职责。
⑥ 作业许可(PTW)相关事件的经验教训。
⑦ 审核❶和评审❷作业许可(PTW)制度的发现。

更新培训通常需要在规定的时段(每一年或两年)内进行，或者在作业许可(PTW)制度发生变更或作业许可(PTW)系统接受过审计之后根据审计建议进行。

- **应建立程序监测和审计作业许可(PTW)系统**

为了确认(工作人员)是否遵守详细的作业指导书以及控制措施是否有效实施，除了由作业许可(PTW)签发人进行的检查核实之外，现场负责人和现场主管人员也应进行作业许可(PTW)的监督检查。通过(作业)许可监督检查获得的信息应用于强化现场安全工作实践。

---

❶在本书中，审核和审计的含义相同，不加区分。
❷在本书中，"review"根据上下文翻译为评审或审查。

The PTW system should be audited regularly, by competent people, preferably external to the site and who are familiar with local management system arrangements. The audit process should examine monitoring records. Non-conformance with PTW system should be recorded, and subsequent remedial measures tracked to ensure all issues are effectively closed out.

The PTW system should be reviewed regularly to assess their effectiveness. This review should include both leading and lagging indicators as well as specific incidents that could relate to inadequate control of work activity.

1.2.3 Related Life-Saving Rules

Life-Saving Rule 7: Work with a valid work permit when required

Life-Saving Rule 8: Verify isolation before work begins and use the specified life protecting equipment

Life-Saving Rule 11: Obtain authorisation before starting excavation activities

Life-Saving Rule 12: Conduct gas tests when required

Life-Saving Rule 18: Obtain authorisation before overriding or disabling safety critical equipment

1.2.4 References

UK HSE, Guidance on permit-to-work systems: A guide for the petroleum, chemical and allied industries, HSG250, First edition, published 2005.

作业许可（PTW）系统应定期由胜任人员进行审计，最好是熟悉现场管理体系要求的作业现场外部人员。审计过程应查看监督检查记录。应记录作业许可（PTW）系统的不符合项，并跟踪后续整改措施，以确保所有审计发现都得到有效关闭。

应定期审查作业许可（PTW）系统以评估其有效性。除了可能跟作业活动控制措施不足相关的具体事件，审查还应包括先导性指标和跟随性指标。

1.2.3 相关保命法则

保命法则 7：高危作业必须持有有效的作业许可

保命法则 8：开始作业之前确认隔离到位，并使用规定的防护装备

保命法则 11：在动土作业前必须获得授权

保命法则 12：按要求进行气体检测

保命法则 18：在停用安全关键设备（系统）之前必须获得授权

1.2.4 参考资料

UK HSE, Guidance on permit-to-work systems: A guide for the petroleum, chemical and allied industries, HSG250, First edition, published 2005.

1.3 Workplace HSE Meetings

1.3.1 Prompts

(1) Is a process to formally and informally discuss and communicate workplace HSE risks in place and operational?

(2) Do HSE meetings allow meaningful communications up, down and laterally throughout the organisation to facilitate and plan for continuous HSE improvement?

(3) Are HSE meetings minuted using a standardised format that is action and result oriented?

(4) Is a procedure to conduct TBT in place and operational?

(5) Are TBT performed for non-routine jobs that are part of the PTW system?

(6) Are TBT performed for routine jobs that have a JHA?

(7) Are TBT performed for routine jobs that do not have a JHA?

(8) Is a procedure to conduct HSE Briefings in place and operational?

(9) Are HSE Briefings conducted as part of Daily Operations Meeting?

(10) Are HSE Briefings conducted as part of shift changeover?

(11) Is a daily meeting about Permit to Work conducted?

(12) Does the daily meeting about Permit to Work formally discuss HSE issues surrounding permits for the day?

1.3 工作场所 HSE 会议

1.3.1 督导提示

(1) 正式和非正式讨论与交流工作场所 HSE 风险的程序是否建立并实施？

(2) HSE 会议是否在整个组织内进行纵向(向上、向下)和横向有意义的沟通以促进和规划 HSE 管理的持续提升？

(3) HSE 会议是否有会议记录？会议记录格式是否使用标准化的格式并以行动和结果为导向？

(4) 召开工具箱会议(TBT)的程序是否建立并实施？

(5) 对非常规作业是否召开工具箱会议(TBT)？该会议是否作为作业许可(PTW)制度的一部分？

(6) 对使用工作危害分析(JHA)的常规作业是否召开工具箱会议(TBT)？

(7) 对不使用工作危害分析(JHA)的常规作业是否召开工具箱会议(TBT)？

(8) 召开 HSE 择要说明(会)的程序是否建立并实施？

(9) HSE 择要说明(会)是否作为每日生产例会的一部分？

(10) HSE 择要说明(会)是否作为换班过程的一部分？

(11) 是否召开关于作业许可(PTW)的每日例会？

(12) 关于作业许可(PTW)的每日例会是否正式讨论当天所有跟作业许可(PTW)相关的 HSE 问题？

1.3.2 Reading

Workplace HSE meetings are a key part of building a HSE culture, raising worker HSE awareness and improving HSE risk management. HSE meetings motivate workers to implement HSE and to move HSE from the HSE - MS documents out into the practice. HSE meetings can be formal or informal and can cover a variety of topics.

Typical workplace HSE meetings include Toolbox Talks(TBT) and the HSE Briefing, where:

① A TBT is a short pre-work talk between the supervisors and personnel discussing the HSE hazards/risks associated with a specific activity such as lifting, driving etc. TBT may also be referred to as a pre-job safety talk, or a pre-job meeting, etc.

② A HSE Briefing is a short briefing by the supervisors to the personnel on the HSE hazards/risks associated with the overall work planned for the shift.

- **A process to formally and informally discuss and communicate workplace HSE risks should be in place and operational**

HSE meetings should be held to allow meaningful communications up, down and laterally throughout the organisation to facilitate and plan for continuous HSE improvement.

All HSE meeting types should encourage active participation in HSE risk management by allowing every employee a right to speak and be heard.

Records should be maintained for all HSE meetings. The minutes of the meeting should contain the following information:

1.3.2 知识准备

工作场所 HSE 会议是培育 HSE 文化、提高员工 HSE 意识和改善 HSE 风险管控的关键部分。HSE 会议激励作业人员践行 HSE(要求),将 HSE(要求)从 HSE 管理体系(HSE-MS)文件落实到工作现场。HSE 会议可以是正式的或非正式的,可以涵盖各种主题。

工作场所典型的 HSE 会议包括工具箱会议(TBT)、HSE 择要说明(会),其中:

① 工具箱会议(TBT)是现场主管人员与员工之间在工作前的简短交谈,旨在讨论与特定活动相关的 HSE 危害及其风险,例如吊装作业、驾驶作业等。工具箱会议(TBT)也被称为作业前安全交谈(谈话、喊话)或作业前会议等。

② HSE 择要说明(会)是由现场主管人员向员工简要介绍与本班作业整体工作计划相关的 HSE 危害及其风险。

- **应建立程序正式或非正式地讨论和沟通工作场所 HSE 风险并有效实施。**

应召开 HSE 会议以便在整个组织机构内进行纵向(向上、向下)以及横向的有意义的沟通,来促进和规划 HSE(管理)持续提升。

所有类型的 HSE 会议都应给予每位员工发言和被人们听到的权利,以鼓励每位员工积极参与 HSE 风险管控。

所有 HSE 会议都应生成会议记录,会议记录应包含以下信息:

① Date and time meeting was held.
② Names of participants.
③ Proposals and subsequent discussion.
④ Action items.
⑤ Nominated person or position responsible for implementing action item(s) with an anticipated closeout date.

- **A procedure to conduct TBT should be in place and operational**

A procedure for controlling TBT should be implemented at the workplace.

A TBT is a short meeting, typically 10-15 minutes duration. It is held among members of the work team to discuss the HSE hazards and their risks related to the specific work activity being performed. TBT may be performed as part of PTW or as part of routine work activities(with or without JHA).

The purpose of a TBT should be to discuss with the working parties:

① What the work activities will be.
② The identified hazards that will require controls during the course of the work.
③ What the controls are and the work parties who will be responsible for their effective implementation.

A TBT as part of PTW for non-routine work activities should be conducted by the Performing Authority before starting any activity under the PTW. The TBT should discuss the work permit, JHA and any supporting certificates, with the persons who will actually be carrying out the work, and therefore be directly exposed to possible hazards.

① 会议召开的日期和时间。
② 参会人员姓名。
③ 提案及其讨论。
④ 行动项。
⑤ 执行行动项的指定负责人或岗位，行动项的预期关闭日期。

- **应建立召开工具箱会议（TBT）的程序并有效实施**

工作场所应实施工具箱会议（TBT）的控制程序。

工具箱会议是一个简短的会议，通常 10~15min。会议在作业班组成员之间召开，用以讨论跟所执行具体作业活动相关的 HSE 危害及其风险。工具箱会议可作为作业许可（PTW）或常规作业活动（使用或不使用工作危害分析）的一部分来实施。

工具箱会议（TBT）的目的是与作业相关方讨论：
① 将执行什么样的作业活动。
② 识别出的需要在作业过程中管控的危害。
③ 采取的防控措施及有效实施防控措施的责任人。

工具箱会议（TBT）作为非常规作业活动作业许可（PTW）的一部分，应在作业活动开始之前由作业许可（PTW）的实施主体组织召开。在召开工具箱会议（TBT）时，应与实际执行该项作业的人员讨论作业许可、工作危害分析（JHA）和任何支持性的专项作业单，因为实际执行作业的人员直接暴露于可能的作业危害中。

Depending upon the nature of the task the Performing Authority may choose to delegate the responsibility of conducting the TBT to a Subject Matter Expert (SME) or a competent person who is the most knowledgeable person in the team.

The TBT shall be recorded on the TBT Record Form, which provides an effective communication means for ensuring the work parties are:

① Aware of the work scope.
② Aware of the hazards.
③ Knowledgeable as to the methods/procedures to be adopted.
④ Aware of the work (or PTW) precautions and controls.
⑤ Aware of any constraints.
⑥ Aware of what tools and equipment are to be used.
⑦ Aware of any environmental considerations.
⑧ Aware of any potentially conflicting activities.
⑨ Able to clarify any issues and provide comments.

A TBT (with JHA) for routine work activities should ensure that hazards and their risks for the job have been identified, effective controls are in place and to make sure that workers involved in a particular job understand those hazard and controls.

A TBT (without JHA) for routine work activities is intended to be a quick and simple risk assessment looking for the day to day hazards which are likely to cause low levels of harm. If significant harm is identified then a JHA should be performed.

The TBT shall be conducted by a Facilitator, who is either the Site Supervisor or the leader of the work team. The Facilitator does not need to be an HSE expert, but shall be qualified and competent to supervise the job from all aspects including HSE and quality of the work.

根据工作性质,(作业)实施主体可授权该领域的专业人员(SME)或班组中知识最丰富的胜任人员主持召开工具箱会议(TBT)。

工具箱会议(TBT)应记录在专门的工具箱会议(TBT)记录表中,该记录表提供了一种有效的沟通方式,以确保作业相关方:

① 清楚作业范围。
② 清楚作业危害。
③ 明白所要采用的(作业)方法/程序。
④ 清楚作业(或作业许可)的(危害)防范和控制措施。
⑤ 清楚约束条件。
⑥ 清楚所要使用的工具和设备。
⑦ 清楚任何作业环境方面的考虑。
⑧ 清楚任何存在潜在冲突的活动。
⑨ 能够澄清任何问题项并提供意见。

常规作业活动[使用工作危害分析(JHA)]的工具箱会议(TBT)应确保识别出作业危害及其风险,并制定出有效的控制措施,同时也应确保参与作业活动的人员了解这些危害及其控制措施。

常规作业活动[不使用工作危害分析(JHA)]的工具箱会议(TBT)本身作为快速简便的风险评估过程,识别可能造成轻度损伤的日常危害。如果识别出重大危害,则应开展工作危害分析(JHA)。

工具箱会议(TBT)应由专人主持召开,主持人可以是现场主管人员或者作业班组的负责人。主持人不需要是HSE专家,但应具备从HSE和工作质量等多方面监督工作的资质和能力。

The TBT requires active participation of each member of the work team and an open exchange of information among the team members. It is the Facilitator's responsibility to lead the discussions and create an environment of openness and good team work.

The TBT should not be a passive discussion or listening exercise. The TBT opportunity is maximized when the entire team actively participates. In an effective TBT, the facilitator will call on the workers and ask them to demonstrate understanding of specific job steps and tools that will be used, and of hazards associated with the tools. Facilitator should also discuss ways that hazards are controlled in the job plan.

Questions about the job are not only welcome but are encouraged and expected. A healthy discussion and clear understanding of the job helps to bring workers back home just as they left—safe and healthy.

- **A procedure to conduct HSE Briefings should be in place and operational**

A procedure for conducting the HSE Briefings should be implemented at the facility.

HSE Briefings are short (10 – 15 minutes) and HSE focused meetings conducted by the Site Supervisors or the HSE Supervisor (as the Facilitator) for the purpose of increasing HSE awareness and reinforcing the need for effective HSE control in work activities among the personnel prior to the start of work. The HSE briefing is normally incorporated into the Daily Operations Meeting to provide a means for discussing risks affecting the site and HSE programmes.

The objectives of HSE Briefings are similar to TBT but they differ in scope and content:

工具箱会议(TBT)需要作业班组中每位成员的主动参与以及成员之间坦率的信息交流。主持人应引导会议讨论并营造开放和良好协作的氛围。

工具箱会议(TBT)不应该是消极的讨论或听力练习。当整个班组积极参与时，工具箱会议(TBT)的效果就会最大化。在有效的工具箱会议(TBT)期间，主持人会号召作业人员描述其对具体作业步骤和所使用工具的理解，以及对所使用工具相关危害的理解。主持人还应(与作业人员)讨论工作计划中的危害控制措施。

(在工具箱会议中)提出与工作有关的问题不仅应受到欢迎，而且应得到鼓励和期待。对工作进行正常的讨论和清晰的理解，使得员工"高高兴兴上班来，平平安安回家去"。

- **应建立召开HSE择要说明(会)的程序并有效实施**

生产作业场所应实施HSE择要说明(会)的控制程序。

HSE择要说明(会)是简短(10~15min)并聚焦于HSE的会议，它在作业开始前由作业现场主管人员或HSE主管(作为主持人)组织召开的，它以提升作业场所人员的HSE意识和强化他们对作业活动有效HSE管控措施的需要为目的。HSE择要说明(会)通常与每天的生产作业例会合并召开，旨在讨论影响作业现场的HSE风险及其管控方案。

HSE择要说明(会)的目的与工具箱会议(TBT)相似，但在范围和内容方面有所不同：

① HSE Briefings, typically do not focus on a specific job, instead they cover all work activities planned for the shift/day. They may also discuss future work (typically within the coming week) where additional awareness or work planning may be required.

② HSE Briefings involve participation of the entire shift personnel including any contractors working within the site/area.

HSE Briefings should also be used as part of shift changeovers.

The Site/Shift Supervisors Logbook should be used as a record for all HSE Briefings.

- **A procedure to conduct daily meetings about Permit to Work should be in place and operational**

A procedure for conducting the daily meeting about permit to work should be implemented at the workplace.

As part of the communication process within the PTW system the site supervisor shall hold a daily meeting with the specific purpose of discussing the status of the current permits (either for closure or revalidation) and to perform a review of all the new permits which may be required for the night shift and the next day. It is expected that this meeting will be held every day as an opportunity to discuss and agree on the work schedule for the night shift and the next day.

This process ensures that all of the relevant personnel are aware of the status for each permit including, as a minimum, Operations, Maintenance, HSE and Contractors. Site HSE personnel shall attend the meeting in an advisory capacity and not as the owner of the PTW system.

① HSE择要说明(会)，通常不聚焦于一项特定作业，而是覆盖当班或当天计划的所有作业活动。会议也可能讨论下一步(通常一周之内)可能需格外注意或需提前计划的作业。

② HSE择要说明(会)的参会人员应包括在作业现场的所有班组成员以及承包商人员。

HSE择要说明(会)也可以作为换班工作的一部分。

作业场所/班组主管的工作日志应记载所有HSE择要说明(会)(的内容)。

- **应建立召开作业许可(PTW)每日例会的程序并有效实施**

作业场所应实施关于作业许可(PTW)每日例会的控制程序。

作为作业许可(PTW)制度沟通过程的一部分，作业场所主管人员应每天召开一次会议，讨论当前所有作业许可(PTW)的状态(关闭或重新生效)，审查夜班或第二天可能需要的新作业许可(PTW)。这类会议通常每天召开，以提供机会讨论和商定夜班和次日的工作时间表。

这个过程确保所有相关人员清楚每个(作业)许可的状态，相关人员至少应包括作业、维修、HSE和承包商人员。现场HSE人员应作为顾问而不是作业许可(PTW)制度负责人参加会议。

1.4 Emergency Management and Crisis Management

1.4.1 Prompts

(1) Have all emergency and crisis situation hazards been identified, risks assessed and used to establish a plan to safely manage emergency and crisis situations?

(2) Have all process related emergencies (e. g. fire, explosion and toxic events) been identified and credible emergency scenarios and response actions developed?

(3) Have all non-process related emergencies (e. g. Security, Aviation, Land Transport, Contagious Disease, Medevac, etc.) been identified and credible emergency scenarios and response actions developed?

(4) Is a tiered Incident Command System (ICS) based on Gold (strategic), Silver (tactical) and Bronze (operational) emergency responses in place and in a state of readiness?

(5) Is a Site Emergency Response Plan (SERP) in place and in a state of readiness?

(6) Does the SERP contain all the process and non-process emergencies identified in the risk assessment?

(7) Have emergency escalation criteria been established to trigger the initiation of Incident Command System response tiers?

(8) Is a process to test SERP and higher level plans in place and operational?

1.4 应急管理和危机管理

1.4.1 督导提示

（1）是否已识别了紧急❶和危机事态下的所有危险？是否对危险进行了风险评估？是否据此制定了安全应对紧急和危机事态的计划？

（2）是否已识别了跟工艺过程相关的所有紧急事态[例如火灾、爆炸和有毒（物质释放）事件]？是否构建了可信的紧急情景及其应对措施？

（3）是否已识别了非工艺过程相关的所有紧急事态（例如安保、航空、陆路运输、传染病、医疗转运等）？是否构建了可信的紧急情景及其应对措施？

（4）是否建立了基于金（战略）、银（战术）和铜（操作）级应急响应的分层突发事件应急指挥系统（ICS）？该系统是否处于就绪状态？

（5）现场应急响应计划（SERP）是否到位并处于就绪状态？

（6）现场应急响应计划（SERP）是否包含了风险评估识别出的所有工艺过程和非工艺过程紧急事态？

（7）是否已建立紧急事态升级准则以触发突发事件应急指挥系统（ICS）不同响应层级的启动？

（8）是否建立了程序用于测试现场应急响应计划（SERP）和更高层级的应急响应计划并有效实施？

---

❶在本书中，"emergency"根据上下文翻译为紧急、紧急事态、紧急状态、紧急状况、紧急情况、紧急事件、突发事件、应急等。

(9) Are full scale major accident hazard (MAH) emergency simulations, involving mutual aid partners performed and documented?

(10) Are regular exercises and drills couducted to test the effectiveness of the Emergency Response Plan (ERP)?

(11) Have mutual aid facilities/services been adequately tested as part of MAH emergency simulation exercises?

(12) Are lessons learned from drills and simulation exercises used to review and update the SERP and critical equipment requirements?

(13) Is emergency communication equipment available, clearly identified and described?

(14) Has the integrity of emergency communication in relation to major accident scenarios been assessed?

(15) Have emergency access/egress and response times for emergency vehicles (Ambulance & Fire Engine) been assessed?

(16) Is an adequately equipped and resourced Emergency Response Centre (ERC) in place and in a state of readiness?

(17) Does the ERC locally maintain adequate emergency response equipment to deal with the most credible accident scenarios identified in the SERP?

(18) Are Medevac facilities available for emergency evacuation of serious MAH casualties?

(19) Is a process to control the inspection, maintenance and testing of emergency response equipment in place and operational?

(9) 是否进行了涉及互助伙伴的完整的重大事故危害(MAH)紧急事态模拟并进行了记录？

(10) 是否定期进行训练和演习以测试应急响应计划(ERP)的有效性？

(11) 作为重大事故危害(MAH)紧急事态模拟训练的一部分，是否对互助设施/服务进行了充分测试？

(12) 是否将演练和模拟训练中得到的经验教训用于审查和更新现场应急响应计划(SERP)和关键设备要求？

(13) 应急通信设备是否可用并清晰地标识和描述？

(14) 是否评估了跟重大事故情景有关的应急通信完整性？

(15) 是否评估了应急车辆(救护车和消防车)的紧急出入口和响应时间？

(16) 是否建立了装备齐全、资源充足的应急响应中心(ERC)并处于就绪状态？

(17) 应急响应中心(ERC)是否在本地持有足够的应急响应装备来处理现场应急响应计划(SERP)中确定的最可信事故情景？

(18) 对重大事故危害(MAH)造成的危重伤员(实施)紧急转运时，医疗转运设施是否可用？

(19) 检查、维护和测试应急响应设备的控制程序是否建立并有效实施？

(20) Is a process in place to ensure that the safety critical personnel who work with emergency and crisis management are competent?

(21) Do all personnel in emergency response positions understand the accountabilities, roles, responsibilities and requirements of their allocated position?

(22) Is a process to monitor, audit and improve the performance of emergency and crisis management in place and operational?

1.4.2 Reading

An emergency is an HSE incident that poses an immediate threat to human life, serious damage to property and assets, the environment, and/or the security of a facility, which requires an immediate response for controlling the situation and the consequences. A Crisis is an incident, emergency or other set of circumstances, which significantly threatens the operations of the company. Incidents can escalate into emergencies which unless controlled can lead to crisis situations.

The Incident Command System is a standardised approach to the command, control, and coordination of emergency response and crisis management, providing a common hierarchy within which responders from multiple agencies can be effective. There should be at least three levels of oversight involved in managing an emergency or crisis event:

① Gold(Strategic): Gold Control is the highest level Company Command Level with overall command and responsibility for formulating the strategy to manage crisis situations. Delegates tactical decisions to Silver Control.

(20) 是否建立了程序确保从事应急和危机管理的安全关键人员胜任？

(21) 应急响应岗位的所有人员是否理解其所分派岗位的问责制、角色、职责和要求？

(22) 是否建立了监测、审核和改进应急和危机管理绩效的程序并有效实施？

1.4.2 知识准备

跟健康、安全、环境(HSE)相关的紧急事件是对人的生命构成直接威胁，(会)对财产和资产、环境和/或设施安全造成严重破坏的 HSE 事件，需要立即响应以控制事态和后果。跟 HSE 相关的危机或危机事件❶是严重威胁公司运营的事件、紧急状况或其他情形。普通 HSE 事件可能演变为紧急事件，如未受控，(普通 HSE 事件)可能进一步演变为危机事态。

突发事件应急指挥系统(ICS)是指挥、控制和协调应急响应和危机管理的标准化方法，提供了多主体有效应急响应的通用层级结构。管理紧急事件或危机事件应至少涉及三个层级的控制：

① 金层级(战略上)：金层级控制是最高的公司❷控制层级，承担总控制和制定危机事态应对策略的责任。授权银层级控制进行战术决策。

---

❶危机事件的英文也翻译为 crisis event。
❷Company 也可称为企业。

② Silver (Tactical): Silver Control is the set-up remote from the scene of operations in a place where communications and relevant information is available. Silver Control will confirm the operational tactics to be deployed to achieve the crisis management strategy set by Gold.

③ Bronze (Operational): Bronze Control will control and deploy the resources within a geographical sector and will implement the tactics as specified by Silver. There may be more than one Bronze Control; each will have an Incident Controller and sufficient support to achieve its operational objectives.

- **A process should be in place to identify hazards created by emergencies and crisis situations, to assess risks and to establish a plan to safely manage emergency and crisis situations**

Emergency and crisis situations should be avoided. Before emergencies and crisis events occur, risk assessment should proactively consider all measures that can be implemented to reduce credible emergency scenarios and risks to a tolerable level and to return the facility to a safe state.

Hazardous events with the potential to cause a Major Accident should be identified as part of an emergency and crisis situations risk assessment:

① Flammable substance releases that can lead to major fires and explosions.

② Toxic substance release that can lead to toxic cloud dispersion over facilities and neighbouring areas.

② 银层级(战术上)：银层级控制是在远离应急操作现场、可获得通信和相关信息的地方设置的控制。银层级控制确认应急操作战术得以部署，以实现金层级控制制定的危机事态应对策略。

③ 铜层级(操作上)：铜层级控制是在一个(应急操作)地理区域内控制和部署(应急)资源，实施银级控制确定的战术。可能有不止一个铜层级控制；每个铜层级控制都有一个事件管制员以及足够的支持来实现其应急操作目标。

- **应建立程序识别紧急和危机事态下的危险，评估其风险并制定安全应对紧急和危机事态的计划**

(首先)应避免紧急和危机事态的发生。在紧急和危机事件发生前，风险评估应主动考虑可以实施的所有措施，目的是将可能发生的紧急事态及其风险降低到可容许的水平，并将设施恢复到安全状态。

应将可能导致重大事故的危险事件确定为紧急事态和危机事态风险评估的一部分：

① 可导致重大火灾和爆炸的易燃物质释放。

② 可导致有毒云团扩散到设施和邻近区域的有毒物质释放。

- **An Emergency Response Plan(ERP) should be in place and ready to respond to site emergencies**

An emergency and crisis management process and organisation must be in place to identify, document and respond to credible HSE emergency incident and escalation scenarios. This requires the development and testing of appropriate emergency management and crisis management plans and procedures to respond to each of the identified incident scenarios and for mitigating the HSE risks that may be associated with them.

The ERP should be developed to respond to a range of emergency scenarios ranging from smaller emergencies (e. g. broken limbs) up to full scale crisis situations (e. g. process releases, fires and explosions, militant attack, etc.) that may require mutual aid support and ongoing assistance from higher levels of the Company's ERP structure (i. e. Gold). An ERP should be simple and straightforward and flexible.

The objectives of an ERP are to:

① Follow tactical directions for crisis situations as set by Gold/Silver Control.

② Contain and control incidents so as to minimise their effects and to limit damage to persons, the environment and property.

③ Detail the emergency response and crisis management measures necessary to protect persons, environment, assets and reputation from the effects of major accidents.

④ Provide practical directions for emergency response and crisis management at that location and in specific emergency situations.

⑤ Communicate the necessary information to employees, contractors, the public, police, civil defence, other relevant government departments, relevant agencies and the Company's Silver Control.

- **应制定应急响应计划(ERP)并随时响应现场突发事件**

应建立应急和危机管理流程及组织以识别、记录和响应可能发生的 HSE 紧急事件及其升级情景。为此,需要开发并测试适当的应急管理和危机管理计划以及程序,以响应每个已识别的(突发)事件情景,并削减可能与之相关的 HSE 风险。

应制定应急响应计划(ERP)以应对各种紧急情况,从较小的紧急事态(例如肢体骨折)到大规模的危机事态(如工艺泄漏、火灾和爆炸、武装袭击等),后者可能需要互助支持和来自公司应急响应计划(ERP)中更高层级(如金层级)的持续支援。应急响应计划(ERP)应简单、明确和灵活。

应急响应计划(ERP)的目标是:

① 遵循金/银控制层级设定的应对危机事态的战术指令。

② 遏制和控制(突发)事件以减少其影响,限制其对人员、环境和财产的损害。

③ 详细描述应急响应和危机管理的必要措施,以保护人员、环境、资产和声誉免受重大事故的影响。

④ 为该作业场所和特定紧急情况下的应急响应和危机管理提供符合实际的指令。

⑤ 向员工、承包商、公众、警察、民防、其他相关政府部门、相关机构和公司银控制层级传达必要的信息。

Each facility or operation should maintain a Bronze level Incident Management Team (IMT) comprised of asset or operating area level personnel responsible for managing the safe and rapid response to incidents and emergency situations occurring at or threatening their assets or operating areas. The IMT must:

① Size up the incident, its potential, and the nature and status of tactical response operations.

② Establish "Command and Control" through the development of an overall strategy and objectives for emergency response operations.

③ Direct response actions of personnel to control and mitigate the emergency.

④ Restore normal operations while minimising impacts to people, property, and the environment.

⑤ Keep the Silver Level crisis management team briefed on the status and nature of the emergency as well as the potential for incident escalation.

An Emergency Control Centre (ECC) provides site based centres/facilities for the purpose of emergency response and crisis management. An ECC may comprise either dedicated facilities or a suitable suite of rooms which can be quickly adapted for use in an emergency. The ECC must be sized and outfitted to support each of the response teams defined in the emergency and crisis organisations and appropriate to the expected level of response. If the ECC is not a dedicated facility, plans must be maintained showing the Centre's location and layout including communications equipment, documents and stationery, so that it can be made operational in a timely manner. The ECC must have the capability to maintain effective communication links with response personnel in the field as well as senior management.

每个设施或每项作业应有一个铜层级事件管理小组（IMT），该小组由设施或作业区域内的人员组成，负责对发生在或威胁所在设施或作业区域的事件和紧急事态进行安全和快速响应。事件管理小组（IMT）必须：

① 迅速对事件作出判断，包括事件的潜在后果以及战术响应操作的性质和状态。

② 通过构建应急响应操作的总体战略和目标，建立（事件的）"指挥与控制"。

③ 指导（现场）人员的响应行动以控制和减轻紧急事态。

④ 在尽可能减少对人员、财产和环境的影响的同时，恢复正常作业。

⑤ 让银层级危机管理小组了解突发事件的状态和性质，也了解事件升级的可能性。

应急控制中心（ECC）提供基于事件发生场所的中枢/设施，其目的是用于应急响应和危机管理。应急控制中心（ECC）可以由专用设施或者由经过快速调整即适用于紧急事态的合适房间组成。应确定应急控制中心（ECC）的大小和装备，以支持紧急和危机组织中定义的不同响应团队，并且适用于各种不同响应级别。如果应急控制中心（ECC）不是应急控制专用设施，则必须维护有效的预案，展示 ECC 的位置和布局，包括通信设备、文件和文具，以便能适时投入使用。应急控制中心（ECC）必须能够与现场响应人员以及公司高层管理人员保持有效的通信联系。

Suitable emergency response and crisis management equipment must be rapidly available and in a state of readiness. Each facility must maintain emergency response and crisis management equipment of a type and quantity that is sufficient to deal with the most probable accident scenarios identified in the ERP. For larger low probability crisis events or offsite escalating crisis events, facilities are expected to enter into Mutual Aid agreements for the provision of additional equipment.

All emergency response equipment under the direct control of a facility must be included in an inspection, maintenance and testing program to ensure it is kept in a constant state of readiness.

- **A process should be in place to ensure that personnel involved with the ERP are competent**

A training programme must be developed to ensure that all individuals and teams in the emergency response and crisis management organisation are competent to complete their assigned duties.

Members of the tiered emergency response and crisis management organisation and relevant support staff must receive training ranging from a basic introduction to the emergency management system to more advanced aspects of emergency response depending on the nature of their job. Refresher courses must be provided at a predetermined frequency for all members of the emergency organisation.

The training program must be modified as required to reflect changes to procedures, responsibilities, and lessons learned from exercises, drills and real emergencies.

恰当的应急响应以及危机管理器材必须能快速处于可用和准备就绪状态。每个作业场所必须保持充足类型和充足数量的应急响应以及危机管理器材，以应对应急响应计划（ERP）中确定的最可能的事故情景。对大规模低概率危机事件或作业场所外不断升级的危机事件，作业场所需要成为互助协议的一部分，以获取额外的(应急响应及危机管理)器材。

作业场所直接控制的所有应急响应器材都应通过专门程序进行检查、维护和测试，以确保其始终处于准备就绪状态。

- **应建立程序确保参与应急响应计划的人员胜任**

应制定培训方案以确保应急响应和危机管理组织中的所有个人和团队都有能力完成分配给他们的职责。

分层应急响应和危机管理组织的成员以及相关支持人员必须接受培训。根据每人的岗位性质，培训从基本的应急管理系统介绍到应急响应更高级的内容。应急组织中的所有成员必须按预设频次接受更新培训。

培训方案必须根据需要不断地更新，以反映应急程序和人员职责的变化以及训练、演习和实战中获得的经验教训。

Drills and exercises must be conducted to assess and improve emergency response/crisis management capabilities, including liaison with and involvement of external organizations.

All elements of the ERP should be routinely tested at least annually. Notification and call-out drills that test communications channels and contactability of key individuals should be routinely conducted at least quarterly.

应进行演习和训练以评估和改进应急响应/危机管理能力,包括联络外部组织及外部组织的参与。

通常,应急响应计划(ERP)的所有要素,至少每年进行一次测试。用于测试通信渠道和关键人员可联系性的通告演习和召集(调遣)演习,应至少每季度进行一次。

2

CONTRIBUTION FACTORS AND ESCALATION FACTORS FOR HSE RISK

HSE风险贡献因素和升级因素

2.1 Fitness to Work

2.1.1 Prompts

(1) Have all health hazards for work activities been identified, risks assessed and used to establish a safe method of work?

(2) Is a Health Risk Assessment (HRA) procedure in place and operational?

(3) Are HRAs performed by competent occupational health professionals or qualified medical practitioners who are experienced in HRA and understand the work activities at operation facilities?

(4) Is a Fitness to Work procedure in place and operational?

(5) Is a Fitness to Work assessment required for activities with significant noise exposures?

(6) Is a Fitness to Work assessment required for confined space entry?

(7) Is a Fitness to Work assessment required for lifting operations?

(8) Is a Fitness to Work assessment required for work at height?

(9) Is a Fitness to Work assessment required for driving (light vehicles, heavy vehicles, cranes, fork lift truck, etc.)?

(10) Is a Fitness to Work assessment required for night working?

(11) Is a Fitness to Work assessment required for lone working?

2.1 健康适岗

2.1.1 督导提示

(1) 是否已识别出作业活动的所有健康危害？是否对其进行了风险评估？是否据此确立了安全的工作方法？

(2) 是否建立了健康风险评估(HRA)程序并有效实施？

(3) 健康风险评估(HRA)是否由具备健康风险评估(HRA)经验、了解生产设施作业活动、胜任的职业健康人员或合格的医疗人员来进行？

(4) 是否建立了健康适岗(评估)程序并有效实施？

(5) 具有显著噪声暴露的活动是否要求开展健康适岗评估？

(6) 进入受限空间是否要求开展健康适岗评估？

(7) 吊装作业是否要求开展健康适岗评估？

(8) 高处作业是否要求开展健康适岗评估？

(9) 驾驶(轻型车辆、重型车辆、吊车和叉车等)是否要求开展健康适岗评估？

(10) 夜间作业是否要求开展健康适岗评估？

(11) 独自作业是否要求开展健康适岗评估？

(12) Is there a health surveillance system to provide periodic worker health assessments in place and operational?

(13) Is there a pre-employment health assessment?

(14) Is the pre-employment health assessment used as a basis to screen out candidates unfit for the role?

(15) Are annual health checks made against Fitness to Work criteria?

(16) Is the annual health check used as a basis to screen out position holders unfit for the role?

(17) Are the Fitness to Work evaluation criteria based on the HRA and used by the health surveillance system that is place and operational?

(18) Are the Fitness to Work evaluation criteria in line with international best practice and Evidence-Based Medicine (EBM)?

(19) Is a process in place to ensure that the medical and occupational health personnel who work with HRA and Fitness to Work are competent?

2.1.2 Reading

"Fitness to Work" assessment is a medical evaluation performed to evaluate whether an employee is medically fit to safely carry out the specific job or task that he/she is employed to do.

This process evaluates individuals against a set of criteria to ensure they are capable of safely undertaking the tasks involved in their job role and are not a risk to themselves or others. The focus of these checks is to enable workers to work safely and productively.

（12）是否建立了对员工进行定期健康评估的健康监护制度并有效实施？

（13）是否进行入职前健康评估？

（14）是否以入职前健康评估为基础筛查（身体健康状况）不适合岗位要求的候选人？

（15）是否根据健康适岗评估准则开展年度健康检查？

（16）是否以年度健康检查为基础筛查（身体健康状况）不适合岗位要求的人员？

（17）是否建立了基于健康风险评估（HRA）并用于健康监护制度的健康适岗评估准则并有效实施？

（18）健康适岗评估准则是否与国际最佳实践和循证医学（EBM）保持一致？

（19）是否建立了程序确保进行健康风险评估（HRA）和健康适岗评估的医疗和职业健康人员是胜任的？

2.1.2　知识准备

"健康适岗"评估是从医学角度评估员工身体健康状况是否适合安全地执行其岗位要求其从事的具体工作或任务。

这个过程是根据一系列准则对人员的身体健康状况进行评估，确保他们有能力安全地执行其岗位要求的任务，并且不会对自己和他人造成伤害。这些检查的焦点是为保证工作人员安全且有成效地工作。

- A process should be in place to identify health hazards for work place activities, to assess risks and to establish a safe method of work

Health Risk Assessments (HRAs) should be performed to identify long and short term health hazards for all work activities. HRAs should be performed by competent occupational health professionals or qualified medical practitioners who are experienced in HRA and understand the work activities at operation facilities.

As a minimum, the following key work activities (deemed to be high health risk potential) should be considered as part of a formally documented HRA:

① Activities with significant exposure to Noise.
② Confined Space Entry.
③ Lifting Operation.
④ Working at Height.
⑤ Driving – Light Vehicles, Heavy Vehicles, Cranes, Fork Lift Trucks.
⑥ Night Time Working.
⑦ Lone Working.

The HRA should consider long term and short term exposure risks to the hazards associated with key work activities. The HRA should identify the control measures that will be required to ensure that health risks are tolerable.

- 应建立程序识别工作场所活动的健康危害，评估其风险并确立安全的工作方法

健康风险评估(HRAs)用于识别所有工作活动的长期和短期健康危害。健康风险评估(HRAs)应由胜任的职业健康人员或有资质的医疗人员来开展，他们应有健康风险评估(HRA)的经验并了解生产设施的作业活动。

至少以下关键作业活动(视为具有潜在的高健康风险)应作为正式的、文件化的健康风险评估(HRA)的一部分：

① 显著暴露于噪声的活动。
② 进入受限空间。
③ 吊装作业。
④ 高处作业。
⑤ 驾驶轻型车辆、重型车辆、吊车和叉车。
⑥ 夜间作业。
⑦ 独自作业。

健康风险评估(HRAs)应考虑与关键作业活动有关的(健康)危害的长期和短期暴露风险。健康风险评估(HRAs)应识别出确保健康风险可容许的控制措施。

- **A Fitness to Work procedure should be in place and operational**

A Fitness to Work procedure should be implemented at site through the occupational health advisory or medical clinician functions. Fitness to Work should be applicable to all employees engaged in key work activities or other high health risk potential work activities as demonstrated by the HRA.

The HRA should be used to determine the fitness to work evaluation criteria. The Fitness to Work procedure should be established accordingly. The Fitness to Work procedure should be aligned with a clearly defined health surveillance programme.

Health surveillance is a risk based system of ongoing health checks required when workers are exposed to hazardous substances or activities that may cause them harm. It helps employers to regularly monitor and check for early signs of work-related ill health in these employees. Health surveillance should be conducted by the Occupational Health Doctor or Occupational Health Nurse.

Health surveillance may be grouped broadly into:

① Biological monitoring, to measure the extent of absorption of a hazardous substance by the employee. This includes biological effect monitoring which is a measure of the extent of biochemical or physical change on an individual after exposure to a hazardous agent.

② Medical evaluation, to detect any adverse effects of a hazard on the employee.

- **应建立健康适岗（评估）程序并有效实施**

健康适岗（评估）程序应在作业现场通过职业健康或医疗人员来实施。健康适岗评估应适用于所有从事关键作业活动或其他健康风险评估（HRA）证明为高潜在健康风险作业活动的工作人员。

健康风险评估（HRA）应用于确定健康适岗评估的准则。健康适岗（评估）程序也应相应建立。健康适岗（评估）程序应与明确制定的健康监护方案保持一致。

健康监护是一项基于风险的制度，该制度需要对暴露于可能导致伤害的危险物质或活动的员工进行适时健康检查。它帮助企业定期监测和检查工作相关疾病的早期症状。健康监护应由职业健康医生或职业健康护士来开展。

健康监护大体上可分为以下组别：

① 生物监测。测量员工吸收有害物质的程度。这包括生物效应监测，即一种测量暴露于有害因素后人员发生生物化学或物理变化程度的方法。

② 医疗评估。检查员工对有害因素的不良反应。

A range of Fitness to Work evaluation criteria should be established for all key work activities. Where there are no specific legal or industry guidelines for the Fitness to Work for a particular job role, then the Company and the occupational health service provider should decide on the content and scope of the criteria and health surveillance. These criteria should be in line with international good practice and be evidence based. An evidence based approach ensures that health evaluation criteria used to make decisions about workers health should be based on best science and driven by internationally recognised authorities such as WHO, UK HSE, OSHA, etc.

Application of the Fitness to Work evaluation criteria to the health surveillance results indicates whether the employee is fit to work, temporary unfit to work or unfit to work.

- **A process should be in place to ensure that personnel who work as part of Fitness to Work are competent**

Occupational Health Doctors should possess skills and expertise including an understanding of the health hazards that can arise at work, the ability to assess risks relating to the health of individuals and groups, knowledge of the local laws relating to workplace issues, and understanding of the way business operates.

2.1.3 References

UK HSE, Occupational health standards in the construction industry, RR584, 2007.

应对所有关键作业活动建立一系列的健康适岗评估准则。如果没有专门的法律或行业指南用于一个特定岗位的健康适岗评估，企业和职业健康服务机构应决定相关准则以及健康监护内容和范围。这些准则应与国际良好实践相一致并有证据支持。基于证据的方法可确保用于判断员工健康的评估准则是基于当前的科学方法并由国际认可的权威机构推动，如世界卫生组织（WHO）、英国健康安全执行局（UK HSE）、美国职业安全与健康管理局（OSHA）等。

将健康适岗评估准则应用于健康监护结果，指出员工健康是否适岗、暂不适岗或不适岗。

- **应建立程序确保从事健康适岗（评估）的人员胜任**

职业健康医生应具备相关技能和专业知识，包括清楚在工作中出现的健康危害，有能力评估个人和群体的健康风险，掌握与工作场所事宜相关的本地法律知识，了解业务运作方式。

2.1.3 参考资料

UK HSE, Occupational health standards in the construction industry, RR584, 2007.

2.2 Adverse Weather

2.2.1 Prompts

(1) Have all adverse weather hazards been identified, risks assessed and used to establish safe methods of work?

(2) Does the site monitor weather conditions and include weather predictions in work planning?

(3) Is a process in place to suspend work activities in the event of adverse weather conditions?

(4) Does the site have adverse weather limits where work is restricted? Do site management understand these limits?

(5) Are work tasks suspended when the weather limits are reached? Can the limits be changed where work is critical or where it is not practical to stop? Is upper management approval required for extensions?

(6) Are adverse weather conditions defined and used in the Manual of Permitted Operations (MOPO)?

(7) Is a process in place for adverse weather crisis management, evacuation and recovery?

(8) Are post-adverse weather checks completed before equipment such as scaffolding and cranes are used? Do these checks include ground stability?

(9) Are buildings well maintained to reduce the likelihood of damage during high winds?

(10) Are there areas of the site where water can pool during heavy rain and create a hazard?

2.2 恶劣天气

2.2.1 督导提示

(1) 是否已识别出所有恶劣天气危害？是否对危害进行了风险评估？是否据此确立了安全的工作方法？

(2) 作业现场是否监测天气状况并在工作计划中包含天气预报？

(3) 是否制定了在恶劣天气情况下暂停作业活动的程序？

(4) 作业现场是否规定了限制作业的恶劣天气极限？现场管理人员是否清楚这些(天气)极限？

(5) 达到(安全作业的)天气极限时，作业任务是否暂停？如果作业任务很关键或者停止作业任务不切实际，可以改变(规定的)天气极限吗？继续作业需要上级管理人员的批准吗？

(6) 许可操作手册(MOPO)中是否定义和使用恶劣天气条件？

(7) 恶劣天气危机管理、撤离和恢复的程序是否建立？

(8) 恶劣天气过后，在使用脚手架和起重机之类的设备前是否完成了检查？这些检查包括地面稳定性吗？

(9) 建筑设施是否得到良好维护以减少因大风受损的可能性？

(10) 作业场所是否有大雨期间聚集雨水并产生危险的地方？

2.2.2 Reading

Adverse weather can be described as any weather state or condition which could have significant impacts upon operations in general, or on a specific work activity. Adverse weather includes high wind speeds (i.e. strong winds, hurricanes), elevated temperatures, excessive rain, flooding, lightning and poor visibility (i.e. fog, mist or sandstorm). The boundary of adverse weather conditions must be defined and applied to operating the facility and all activities associated with supporting the facility operations (e.g. lifting operations, vehicle driving, etc.).

- **A process should be in place to identify adverse weather hazards, assess risks and establish a safe method of work.**

Risk assessments should be performed for a broad range of adverse weather conditions and business activities, including but not limited to lifting operations, scaffolding, driving, aviation operations, etc.

The risk assessment should be used to determine the safe operating envelope for adverse weather conditions. This will involve establishing upper and lower weather criteria for safe operations.

The initial adverse weather criteria should be aligned with the Basis of Design for the facility. The adverse weather risk assessment, criteria and safe operating envelope should be periodically reviewed and updated.

2.2.2 知识准备

恶劣天气可被描述为任何可能对作业普遍产生重大影响或(仅)对某项特定作业活动产生重大影响的天气状况或天气条件。恶劣天气包括高风速(如大风、飓风)、高温、过量降雨、洪水、闪电和低能见度(如浓雾、薄雾或沙尘暴)。应定义恶劣天气条件的界限,并应用于设施运行以及支持设施运行所有相关活动(例如吊装作业、车辆驾驶等)。

- **应建立程序识别恶劣天气带来的危害、评估其风险并确立安全的工作方法**

应针对各种恶劣天气条件和业务活动进行风险评估,业务活动包括但不限于吊装作业、脚手架(作业)、驾驶、航空作业等。

使用风险评估确定恶劣天气条件下的安全操作范围,包括确定安全操作的天气准则上限和下限。

最初的恶劣天气准则应与设施的设计基准保持一致。恶劣天气的风险评估、准则和安全操作范围应进行定期审查和更新。

- **A process should be in place to suspend production and other work activities in the event of adverse weather conditions**

Adverse weather related criteria should be defined in the Manual of Permitted Operations(MOPO). Production and/or other work activities may not be permitted under certain adverse weather conditions. These adverse weather conditions should be clearly established in a MOPO which is normally available in the Central Control Room. The MOPO should include reference to non-production activities including but not limited to land transport, rig movements and aviation operations.

Site management must understand the importance of weather criteria and be prepared to suspend work if the criteria are met. Any override of adverse weather criteria should only be authorised by Senior Management with any deviation supported by a risk assessment with appropriate control measures.

Systems should be in place and operational for weather monitoring and forecasting.

- **A process should be in place for crisis and emergency response, evacuation and recovery**

In the event of adverse weather, emergency intervention criteria should be established for activation of the Site Emergency Response Plan(SERP) under the company Crisis and Emergency Response Plan. The Crisis and Emergency Response Plan should describe adverse weather scenarios and the response to escalating weather conditions, including evacuation of the facility.

- **应建立恶劣天气情况下暂停生产(设施运行)及其他作业活动的程序**

恶劣天气相关准则应在许可操作手册(MOPO)中定义。在某些特定恶劣天气条件下,可能不允许开展生产(设施运行)和/或其他作业活动。这些特定恶劣天气条件应在许可操作手册(MOPO)中明确规定,许可操作手册(MOPO)应放置在中心控制室随时待查。许可操作手册(MOPO)也应提及非生产(设施)运行活动,包括但不限于陆路运输,井队搬家和航空作业。

现场管理人员必须清楚(恶劣)天气准则的重要性,一旦满足恶劣天气准则应暂停相关作业。任何逾越恶劣天气准则的偏离必须基于风险评估并实施合适控制措施,才可以得到管理层的授权。

应建立天气监测和预报制度并有效运行。

- **应建立危机与应急响应、撤离❶及恢复程序**

如果发生恶劣天气,紧急干预准则应予以确立以启动企业危机和应急响应计划下的现场应急响应计划(SERP)。危机和应急响应计划应明确恶劣天气情景以及对不断加剧的恶劣天气条件的响应,包括撤离设施。

---

❶Evacuation 也可翻译为疏散。

Adverse weather can cause damage to equipment, buildings and the ground where the equipment stands. Standing water can cause land destabilisation as well as cause problems for people and traffic.

Emergency recovery and Business Continuity Plans should be established to ensure that the Company can safely resume production activities following cessation of adverse weather event.

Look for evidence that a system is in place to survey vulnerable equipment, buildings and land areas after heavy rain, high winds, etc. to ensure the equipment is safe to use, buildings are not damaged and that the land they sit on is stable and will not collapse. Also look for evidence that large areas of standing water are routinely removed.

2.3 Night Working

2.3.1 Prompts

(1) Have all night working hazards been identified, risks assessed and used to establish safe methods of work?

(2) Is the night working procedure in place and operational?

(3) Are night time workers encouraged to take short regular breaks?

(4) Are fatigued workers removed from the job as a part of normal procedure?

(5) Are rest rooms available for fatigued workers removed from the job?

恶劣天气会对设备、建筑物和设备所在的地面造成破坏。积水可能导致地面松软，并对人员和交通造成问题。

应建立应急恢复和业务连续性计划，以确保公司在恶劣天气停止后安全地恢复生产活动。

查找证据判断以下制度是否到位：暴雨和强风等过后，对易损设备、建筑物和地面进行调查，以确保设备可以安全使用、建筑物没有损坏、设备和建筑物所在区域的地面稳定且不会塌陷。同时，查找证据判断大面积的积水是否被定期清理。

2.3 夜间作业

2.3.1 督导提示

（1）是否已识别出所有与夜间作业有关的危害？是否对危害进行了风险评估？是否据此确立了安全的工作方法？

（2）是否建立了夜间作业程序并有效实施？

（3）是否鼓励夜班作业人员短暂地规律性休息？

（4）从工作岗位撤换处于疲劳状态的人员是否是正常程序的一部分？

（5）是否为因疲劳从工作岗位撤换的人员提供了休息室？

(6) Are warning lights/reflective surfaces deployed around any hazardous area at night (such as around excavations, crane areas)?

(7) Are requirements for shift extensions into night hours included in the PTW system?

(8) Where jobs run longer than expected, are they suspended until the following day where practicable?

(9) Is there a process to control lone working at night in place and operational?

(10) Are sleep and rest patterns for night workers monitored and controlled?

(11) Is a night working training programme in place and operational?

(12) Are permanent night workers put at risk by missing day time training only?

2.3.2 Reading

Working unusual hours has a major effect on mental and physical health. The rhythmical processes that comprise the body's internal clock are coordinated so as to allow for high activity during the day and low activity at night. Working night shifts upsets these rhythms and leads to increased fatigue, stress, and lack of concentration.

Night working increases the likelihood of errors and accidents at work and might have a negative effect on health. Night working introduces an elevated level of safety and operating risk that must be managed.

（6）夜间作业时，是否在任何危险区域周边设置了警示灯/反光面(例如动土作业和起重作业周边)？

（7）作业许可(PTW)制度是否包括延长作业至夜间的要求？

（8）当工作耗时超出预期，是否暂停直到第二天可行时再执行？

（9）是否建立了夜间独自作业的控制程序并有效实施？

（10）是否对夜间作业人员的睡眠和休息模式进行了监测和控制？

（11）是否建立了夜间作业培训方案并有效实施？

（12）错过仅在日间(进行)的培训是否会将长期夜间作业人员置于风险之中？

2.3.2　知识准备

工作时间不正常对身心健康有重大影响。呈节律特征的过程构成人体的内部生物钟，上述过程协调一致以适应白天的大活动量和夜间的小活动量。夜间作业打乱了人体内部生物钟，导致疲劳增加、压力增大和注意力不集中。

夜间作业会增加工作中出现失误和事故的可能性，同时也可能对健康产生负面影响。夜间作业引起安全风险和操作风险升高，这必须加以管控。

- **A process should be in place to identify night working hazards, assess risks and establish a safe method of work**

Risk assessments should be performed for night working for a wide range of routine work activities that are or could be performed at night. Night workers should be involved in the risk assessment.

Night working is a critical part of operations that cannot be totally eliminated. Night working should be avoided where practicable for non-routine, high risk potential, highly complex tasks. When night working is unavoidable, the risk assessment should consider all measures that can be implemented to reduce the risk to a tolerable level.

The following measures should be considered as part of the risk assessment:

① The site and work place should be well lit at night to ensure high visibility and avoid working in the dark.

② Night workers should be allowed to have more frequent short breaks throughout their shift.

③ Ensure, as far as possible, that there is a quiet, secluded area designated for rest and recuperation.

④ Night workers should be encouraged to stay hydrated.

⑤ Night workers should wear the correct PPE, i.e. high visibility clothing, etc.

Managers and supervisors should learn to recognize signs and symptoms of the potential health effects associated with night shifts. Workers who are being asked to work night shifts should be diligently monitored for the signs and symptoms of fatigue. Any employee showing such signs or symptoms should be evaluated and possibly directed to leave the active area and seek rest.

- **应建立程序识别夜间作业危害，评估其风险并确立安全的工作方法**

在夜间开展或可能在夜间开展的各种常规作业活动，应进行风险评估。夜间作业人员应参与到（夜间作业）风险评估中来。

夜间作业是生产作业非常关键的部分，很难完全避免。如果可行，应避免在夜间执行非常规、高潜在风险以及高度复杂的作业任务。如果难以避免，风险评估应考虑所有可能实施的控制措施，从而将风险降低到可容许的水平。

作为风险评估的一部分，以下措施应加以考虑：

① 作业现场和工作场所在夜间应照明到位，以确保高可见度并避免在黑暗中工作。

② 应允许夜班作业人员在整个班次中有更频繁的短暂休息。

③ 尽可能确保有一个僻静不受打扰的区域专门用于休息和恢复。

④ 应鼓励夜间作业人员保持足量饮水。

⑤ 应穿戴正确的个体防护装备，如高亮工服等。

管理人员和现场主管应学会识别与夜间作业相关的潜在健康影响的迹象和症状，并仔细监测夜间作业人员的疲劳迹象和疲劳症状。如有员工显示出相关迹象和症状，应进行评估；如需要，指导该员工离开作业区域休息一下。

Lone working at night should be avoided. If unavoidable, lone working should be supported by a risk assessment with additional controls provided (e.g. regular check-in to the control room, multiple forms of communication, etc.).

Night workers should fully understand how to reduce the risk of fatigue and prepare themselves for a night shift. Measures may include:

① Taking a nap of 1-4 hours before the first night shift.

② Keeping a regular sleeping pattern.

③ Having the largest meal after day time sleep, before starting the night shift.

④ Taking short breaks during the shift.

⑤ Eating balanced and regular meals.

⑥ Avoiding fatty foods entirely during your shift.

⑦ Wearing the correct high visibility PPE.

In order to reduce fatigue risk, camp management is advised to establish day time sleeping arrangements for night time workers so that elevated noise levels during the day do not disturb sleep patterns.

Vehicles colliding with unilluminated obstructions at night can cause fatality and injury to vehicle occupants, damage to the vehicle and the object impacted. Permanent and temporary obstructions (e.g. excavation works) to vehicles should be well illuminated at night or provided with reflective surfaces.

- **A night working procedure should be established and operational**

A procedure should be established to control night time working. The procedure should be used as the basis to develop work instructions, containing site rules for working at night.

应避免在夜间独自作业。如不可避免，夜间独自作业应通过风险评估及额外防控措施加以支持(例如：定期向控制室人员报到，多种形式的联络等)。

夜间作业人员应充分了解如何降低疲劳风险并为夜间作业做好准备。可能的措施包括：

① 在开始第一个夜班作业前小睡 1~4h。
② 保持有规律的睡眠模式。
③ 白天睡眠后，夜班开始前，进食最大餐量。
④ 在夜班期间短暂休息。
⑤ 饮食平衡并且规律。
⑥ 在整个夜班期间避免油腻食物。
⑦ 穿戴正确的、高可见度的个体防护装备。

为减少疲劳带来的风险，建议后勤部门为夜班人员做好日间睡眠安排，以便白天较高水平的噪声不干扰他们的睡眠模式。

在夜间，车辆与未照亮的障碍物碰撞可能导致车上人员的伤亡、车辆受损和被撞击物破坏。车辆周边永久性和临时性障碍物(例如动土作业)应在夜间被充分照亮或设有反光面。

- **应建立夜间作业程序并有效实施**

应建立夜间作业控制程序。该程序应作为制定夜间作业指导书的基准，后者包含夜间作业的现场规定。

It is well accepted that work can be performed better and safer during the day. As such, work planning activities should always consider major work to take place during the day, and where possible, end the same day.

The PTW system should not normally be operational during the night. Exceptions may occur in the case of extending or changing a shift into the night. If a night time job is unavoidable, then a risk assessment and additional control measures, through a formal JHA, should be required to justify night working. Qualified first aiders and medical staff should be on duty and nearby during all night work.

For routine activities that do not require a PTW it is normal for more controls to be required at night due to low surrounding light levels, reduced manpower and fatigue. These additional controls for night working should be captured in the night time JHA for the task.

- **A training programme should be established to build competences for night time workers**

Employees at risk from night working should have received awareness training on the risks and their management.

Permanent or semi-permanent night workers should not be omitted from mandatory HSE training that are normally conducted during the day.

The awareness of workers and their supervisors to night working hazards and risk management should be checked.

2.3.3 References

UK HSE, Managing Shiftwork Health and Safety Guidance, HSG256, (Published 2006).

白天工作更安全更有效，这是广为认可的(观点)。因此，制定工作计划时，应尽量将主要任务安排在白天进行；如果可能，尽量当天完成。

通常，在夜间不开展与作业许可(PTW)相关的作业。只有在作业任务延长或变更到了夜班时可能会出现例外。如果无法避免此类情况，则应通过正式的工作危害分析(JHA)进行风险评估和增加防控措施，以判断作业能否在夜间继续开展。在夜间进行与作业许可(PTW)相关的作业时，有资质的急救人员或医务人员应在附近全程值班。

对于不需要作业许可(PTW)控制的日常作业活动，如果在夜间开展，也需要增加防控措施，这是因为夜间作业存在光照水平低、人员少和疲劳等问题。这些额外的防控措施应通过夜间作业开始前的工作危害分析(JHA)进行识别和确认。

- **应建立培训方案以确保夜间作业人员胜任**

夜间作业人员应接受夜间作业意识培训，清楚夜间作业风险及其防控措施。

由于强制性 HSE 培训通常都在白天进行，(培训计划)应避免遗漏长期或大多数时间从事夜间作业的人员。

应检查作业人员及现场主管人员对夜间作业危害及风险管控的掌握情况。

2.3.3 参考资料

UK HSE, Managing Shiftwork Health and Safety Guidance, HSG256, (Published 2006).

2.4 Work Alone

2.4.1 Prompts

(1) Is a process in place to identify working alone hazards, assess risks and establish a safe method of work?

(2) Have working alone hazards been identified, risks assessed and safe methods of work developed?

(3) Does working alone in a confined space require a buddy system to rescue lone workers?

(4) Does working alone in sour service locations require a buddy system to rescue lone workers?

(5) Is a working alone procedure in place and operational?

(6) Do supervisors periodically visit and observe people working alone?

(7) Is a system in place to ensure regular, pre-agreed communication between supervisor and lone worker?

(8) Is information regarding emergency procedures given to lone workers?

(9) Do lone workers have access to adequate first-aid facilities?

(10) Is a robust system in place to ensure a lone worker has returned to their base or home once their task is completed?

(11) Do employees at risk from working alone receive awareness training on the risks?

2.4 独自作业

2.4.1 督导提示

(1) 是否建立了程序以识别独自作业的危害、评估(其)风险并确立安全的工作方法?

(2) 是否已识别出独自作业的危害?是否对危害进行了风险评估?是否(据此)确立了安全的工作方法?

(3) 在受限空间内独自作业时,是否需要遵循"两人同行制"以(在需要时)救援独自作业人员?

(4) 在含硫环境中独自作业时,是否需要遵循"两人同行制"以(在需要时)救援独自作业人员?

(5) 是否建立了独自作业的程序并有效实施?

(6) 主管人员是否定期探访和观察独自作业的人员?

(7) 是否建立了制度来确保主管人员和独自作业的人员之间进行定期或事先约定的沟通?

(8) 是否向独自作业人员提供了应急程序的相关信息?

(9) 独自作业人员是否可以获得充足的急救设施?

(10) 是否建立了健全的制度以确保独自作业的人员完成作业后返回营地或家中?

(11) 独自作业的人员是否接受了独自作业风险的意识培训?

(12) Has a training programme been established for lone workers?

(13) Is the awareness of workers and their supervisors to the hazards and risks of working alone periodically checked?

2.4.2 Reading

People are "alone" at work when they are on their own or when they cannot be seen or heard by another person. Lone workers include all employees who may go for a period of time where they do not have direct contact with a co-worker.

Oil and gas workers are often required to travel and work in remote locations for a range of activities and are not always accompanied. Lone workers include those that:

① Work remotely or away from a fixed base——for example, maintenance workers, inspectors and those working in enclosed spaces.

② Work separately from others on the same premises——for example, security staff or work outside normal hours.

③ Work at home.

④ Mobile workers——for example, drivers involved in a lot of short or long distance travel.

Not all jobs are suitable for working alone. Working alone introduces an elevated level of health, safety and operating risk that must be managed.

(12) 是否已为独自作业的人员制定了培训方案?

(13) 是否定期检查独自作业的人员及其主管对独自作业的危险及其风险的认识?

2.4.2 知识准备

人们自己工作或工作时不能被他人看到或听到,均为"独自"作业。独自作业的人员包括了所有可能在一段时间内与同事没有直接联系的人员。

石油天然气行业的工作人员经常需要到偏远地点执行一系列(作业)活动,并且不总是有伙伴同行。独自作业的人员包括:

① 在偏远地点或远离固定的营地作业——例如:维修人员、检查人员和在封闭空间作业的人员。

② 在同一处所内,但与其他人员分开工作——例如:安保人员或在正常工作时间以外工作的(人员)。

③ 在家工作。

④ 流动作业人员——例如:从事大量短途或长途驾驶的司机。

并不是所有工作都适合独自作业。独自作业导致健康、安全和操作方面的风险增大,必须有效管控独自作业风险。

Fixed-site workers often work in high risk environments with little or no supervision that can increase the likelihood of errors and accidents at work. This kind of environment poses a threat to lone workers should they experience any difficulties whereby they need assistance. Working alone increases the risk of not being attended to in the event of a medical emergency.

- **A process should be in place to identify working alone hazards, assess risks and establish a safe method of work**

Employees working alone should at all times be at no greater risk than other employees. It is an employer's duty to assess risks to employees and take steps to avoid or control the risks.

Risk assessments should be performed for all work activities where working alone is anticipated or planned. Lone workers should be involved in the risk assessment.

Working alone is a part of oil and gas operations that cannot be totally eliminated. Working alone should be avoided where practicable for non-routine, high risk potential, highly complex tasks. When working alone is unavoidable, the risk assessment should consider all measures that can be implemented to reduce the risk to a tolerable level.

The main risks associated with working alone are:

① Sudden illness or accidents. There is every chance that a lone worker could become ill during their working day or have an accident and be unable to summon assistance.

② Contributing to the loss of a control that can lead to escalation of a risk.

③ Driving related incidents.

固定场所工作人员经常在很少或几乎没有监督的高风险环境中作业，这会增加出错或发生事故的可能性。如果独自作业的人员遇到任何需要援助的困难时，这种环境会对他们造成威胁。独自作业增加了(独自作业人员)在发生医疗紧急情况时不能被及时发现和及时救治的风险。

- **应建立程序识别独自作业的危害，评估其风险并确立安全的工作方法**

在任何时候，独自作业的人员都不应比其他人员面临的风险更大。对员工面临的风险进行评估并采取措施避免或管控风险是企业的职责。

应对预期或计划独自作业的所有活动进行风险评估。进行独自作业的人员应参与该风险评估。

独自作业是石油和天然气行业运营活动的一部分，不能完全消除。对于非常规、高风险、高度复杂的任务，在可行的情况下应避免独自作业。如果独自作业不可避免，风险评估应考虑所有可以实施的措施，以将风险降低到可容许的水平。

与独自作业相关的主要风险是：

① 突发疾病或事故。独自作业的人员很有可能在工作期间生病或发生事故却无法召唤到救援。

② (作为)管控失败的贡献因素，导致风险升级。

③ 发生跟驾驶相关的事件。

These risks are not exhaustive but act as an indication of the risks posed across different roles. A risk assessment provides a more detailed look at the specific risks, enabling measures to be developed to protect lone workers.

Risk assessment should help decide on whether a supervisor may need to be present along with someone dedicated to a rescue role (i.e. a buddy system). There are some high risk activities where at least one other person may need to be present. Examples include:

① Working at a remote location.
② Working with machinery.
③ Working in a confined space.
④ Working in a sour service environment.
⑤ Working in a high security environment.

Employees with certain medical conditions may not be able to work alone on specific work activities. The risk assessment should consider both routine work and foreseeable emergencies that may impose additional physical and mental burdens on a lone worker. Employers should seek medical advice if necessary.

- **A lone working procedure should be established and operational**

A procedure should be established to control lone working. The procedure should be used as the basis to develop work instructions containing site rules for lone working.

Procedures should be put in place to monitor lone workers as effective means of communication are essential. These may include:

① Supervisors periodically visiting and observing people working alone.

② Pre-agreed intervals of regular contact between the lone worker and supervisor, using phones, radios or email, bearing in mind that not all people speak the same language.

以上并非独自作业相关的所有风险，但可作为不同（独自作业）任务所带来风险的示例。风险评估可以更详细地了解具体风险，从而制定具体措施保护独自作业的人员。

风险评估应帮助主管确定是否需要陪同去执行救援任务的人员（如两人同行制❶）。一些高风险活动，至少应有一位陪同人员在场。例如：

① 在偏远地点作业。
② 使用机动设备。
③ 在受限空间中作业。
④ 在含硫环境中工作。
⑤ 在安保风险高的环境中工作。

有某些健康状况的员工，可能无法独自从事特定作业活动。风险评估应考虑某些常规作业以及可预见的紧急情况，它们可能给独自作业的人员带来额外身心负担。如有必要，企业应寻求医疗建议。

- **应建立并实施独自作业程序**

应建立独自作业的控制程序。该程序应作为制定作业指导书的基准，后者包含独自作业的现场规定。

应建立程序监测独自作业人员的情况，因为有效的沟通方式至关重要。这些程序可能包含：

① 主管定期探访和观察独自作业的人员。
② 独自作业的人员和主管之间通过电话、对讲机或电子邮件按既定频次定期联系，应注意并非所有人都说同种语言。

---

❶也称伙伴同行制。

③ Manually operated or automatic warning devices that will trigger if specific signals are not received periodically from the lone worker.

④ Implementing a robust system to ensure a lone worker has returned to their base or home once their task is completed.

Information regarding emergency procedures should be given to lone workers. The risk assessment may indicate that mobile workers should carry first-aid kits. Lone workers should also have access to adequate first-aid facilities.

- **A training programme should be established to build competence for lone workers**

Training is particularly important where there is limited supervision to control, guide and help in uncertain situations. Lone workers are unable to ask more experienced colleagues for help, so extra training may be appropriate. They need to be sufficiently experienced and fully understand the risks and precautions involved in their work and the location that they work in.

Employees at risk from working alone should have received awareness training on the risks and their management.

Employers should set the limits to what can and cannot be done while working alone. They should ensure workers are competent to deal with the requirements of the job and are able to recognise when to seek advice from elsewhere.

The awareness of workers and their supervisors to the hazards and risks of working alone should be checked.

2.4.3 References

UK HSE, Working Alone, INDG73(rev3), Published 05/13, 2013.

③ 如果没有定期从独自作业的人员那里接收到特定信号，手动或自动报警装置则会触发。

④ 实施健全的制度，以确保独自作业的人员在完成作业后返回营地或家中。

应向独自作业的人员提供应急程序的相关信息。风险评估可能会指出，流动作业人员应携带急救箱。独自作业的人员也应该可以获得配备充足的急救设施。

- **应建立培训方案以构建独自作业人员的能力**

如果在具有不确定性的情景中，主管人员的管控、引导和帮助是有限的，培训尤为重要。独自作业的人员无法向更有经验的同事寻求帮助，因此额外的培训可能是恰当的。他们需要有足够的经验并充分了解他们的作业和作业地点所涉及的风险和防范措施。

独自作业的人员应接受过独自作业的风险及风险管控的意识培训。

企业应该明确独自作业时能做什么和不能做什么的界限。他们应确保工作人员胜任岗位要求并能够识别何时从别处寻求建议。

应检查作业人员及其主管人员对独自作业的危险及其风险的掌握情况。

2.4.3 参考资料

UK HSE, Working Alone, INDG73(rev3), Published 05/13, 2013.

2.5 Management of Change

2.5.1 Prompts

(1) Are the change hazards identified, risks assessed and used to establish a plan to manage change safely?

(2) Is a Management of Change (MoC) procedure covering temporary/permanent changes to plant, people and procedures in place and operational?

(3) Is a process to review safety, engineering and technical risks as part of MoC requests in place and operational?

(4) Are all upstream and downstream effected parts of the plant considered during risk assessment of the MoC Request?

(5) Is a process to perform a Mini-HAZOP study and other technical safety studies as part of MoC requests in place and operational?

(6) Is a Mini-HAZOP study performed routinely as part of the safety and technical review of proposed change?

(7) Are the facility's Safety Case and associated registers redlined and noted to keep a record of the proposed change and its impact on process safety of the facility?

(8) Is the facility's QRA used to support the safety engineering of the change, and to ensure that change safety risks are tolerable and that the total safety risks are ALARP?

(9) Is an MoC Register used to monitor the full details and history of temporary and permanent changes in place and operational?

2.5 变更管理

2.5.1 督导提示

(1) 是否识别了变更的危害？是否对其进行了风险评估？是否据此制定了计划以安全地管理变更？

(2) 是否建立了变更管理(MoC)程序并有效实施？该程序是否涵盖了设备、人员和程序的临时/永久变更？

(3) 是否建立了程序将审查安全、设计及技术风险作为变更管理(MoC)申请的组成部分并有效实施？

(4) 在对变更管理(MoC)申请进行风险评估期间，是否考虑了设备上游和下游受到影响的所有部位？

(5) 是否建立了程序将执行微型危险与可操作性(Mini-HAZOP)分析及其他技术安全分析作为变更管理(MoC)申请的组成部分并有效实施？

(6) 微型危险与可操作性(Mini-HAZOP)分析是否作为拟议变更安全审查与技术审查的常规内容加以执行？

(7) 设施的安全例证和相关登记册是否经过标注和注释以记录拟议变更及其对设施工艺安全的影响？

(8) 设施的定量风险分析(QRA)是否用来支持变更的安全设计？是否以此保证变更的安全风险可容许且总体安全风险合理可行尽可能低(ALARP)？

(9) 是否建立了变更管理(MoC)登记册用以监控临时和永久变更的全部细节和历史并有效维护？

(10) Is a process in place to ensure that the safety critical personnel who work with MoC are competent?

(11) Is the MoC Register used to monitor the cumulative HSE risks of numerous smaller changes?

(12) Is a process to monitor, audit and improve the performance of MoC in place and operational?

2.5.2 Reading

Management of Change(MoC) is required to provide assurance that, when changes are made, new risks are not unknowingly introduced, or the prevailing risk profile is not adversely changed without appropriate mitigation.

The MoC process covers the full asset life cycle, involving the functional disciplines that are potentially impacted by the change. It is critical in meeting the business objectives: Asset Integrity, HSE, Production, Quality and Cost. The MoC process ensures that business risks related to change (especially HSE) are identified, mitigated and accepted prior to a change being realised.

MoC may involve permanent, temporary or emergency changes to plant and equipment, procedures, people and organisation, materials and substances.

- **A process should be in place to identify hazards created by change proposals, to assess risks and to establish a plan to safely manage change**

Where changes are made to systems (design changes, operational changes, etc.) the effects of the change should be risk assessed. The risk assessment should consider all measures that can be implemented to reduce any identified risks to tolerable levels.

（10）是否建立了程序确保变更管理（MoC）相关的安全关键人员胜任？

（11）是否使用变更管理（MoC）登记册来监控大量较小变更累积的 HSE 风险？

（12）是否建立了监测、审计及提升变更管理（MoC）绩效的程序并有效实施？

2.5.2　知识准备

为确保进行变更时，不会在不知情的情况下引入新风险，或者在没有适当风险削减措施的情况下，不会对占主导地位的风险状况产生不利影响，需要进行变更管理（MoC）。

变更管理（MoC）程序涵盖整个资产生命周期，涉及可能受变更影响的（每一个）功能领域。它对于实现生产经营目标至关重要：资产完整性、HSE、生产、质量和成本。变更管理（MoC）程序确保在发生变更之前识别、削减和接受与变更相关的生产经营风险（尤其是 HSE）。

变更管理（MoC）可能涉及对设施和设备、程序、人员和组织、材料和物质的永久、临时或紧急变更。

- **应建立程序以识别变更方案产生的危害，评估其风险并制定计划安全地管理变更**

对系统进行变更（设计变更、操作变更等），则应对该处变更带来的影响进行风险评估。风险评估应考虑所有可以实施的措施，将已识别的风险降低到可容许的水平。

Good industry practice requires that process and plant changes should not be undertaken without having undertaken a safety, engineering and technical review. A technical review considers technical and operating integrity risks that share a common basis with safety and engineering but are also distinctly different. For example, the selection of a corrosion inhibitor chemical is a distinct task for the technical integrity and operating functions with implications for safety and engineering. This review should be traceable and identify changes proposed to the following factors:

① Process conditions.

② Operating methods.

③ Engineering methods.

④ Safety.

⑤ Environmental conditions.

⑥ Engineering hardware and design.

A risk assessment should be performed to identify what hazards have been created by the change that may affect plant or personnel safety, and what control measures can be implemented to eliminate or reduce the risk. The risk assessment should be performed by competent personnel who are experienced in risk assessment and the management of change at oil and gas facilities.

Changes may affect other parts of the plant which may be quite remote from the source of the change. Therefore, all parts of the plant should be considered when undertaking hazard identification and risk assessments.

行业良好惯例要求，在未进行安全、设计和技术审查的情况下不应进行工艺和设施变更。技术审查考虑了技术和操作完整性风险，这些风险与安全和设计有着共同的基础，但也有明显的不同。例如，腐蚀抑制化学品的选择是为实现技术完整性和操作功能的独特任务，对安全和设计也有着影响。审查应该是可追溯的，并确定针对以下因素提出的变更：

① 工艺条件。
② 操作方法。
③ 设计方法。
④ 安全。
⑤ 环境条件。
⑥ 工程硬件和设计。

应进行风险评估以确定变更产生的危害对设备或人员安全的可能影响，以及可采取哪些控制措施来消除或降低风险。风险评估应由具备风险评估经验，以及具有石油和天然气设施变更管理经验的胜任人员执行。

变更也可能会影响工厂❶的其他部分，这些部分可能远离变更的源头。因此，在进行危害识别和风险评估时应考虑工厂的所有部分。

---

❶根据情况，plant 翻译为工厂、设施、设备。

A Change HAZOP study should be conducted for the proposed plant change. The objective of the Change HAZOP study should be to evaluate and assess the hazard and operability risks arising from the change. Upstream and downstream risks should be assessed, with safeguarding devices (including procedural controls) determined to mitigate the risks of the change.

Technical safety studies (e. g. HAZID, QRA, SIL, LOPA, bowties, etc.) should be reviewed and updated to assess risks from the proposed change. Upstream and downstream risks should be assessed and risk mitigations determined.

- **An MoC process should be in place and operational**

All changes, whether permanent, temporary or emergency; involving procedures, plant and equipment, people or substances should be subject to a formal management process.

Management should endorse an MoC process that is applied to a range of changes that can impact HSE. The MoC process clearly details the procedures that should be used to manage different types of change. Those personnel in accountable and responsible safety critical positions for MoC shall demonstrate an understanding of the necessity to follow the MoC process and actively participate in the MoC process.

A complete MoC process includes procedures for permanent and temporary changes to plant, procedures, people and organisation. Personnel should be able to demonstrate that they know how to recognise a change requiring an MoC and how to initiate the MoC process. There should be an active process in place to allow personnel to recommend upgrades of the MoC process.

应针对拟议的设备变更进行变更危险与可操作性(HAZOP)分析。变更危险与可操作性(HAZOP)分析的目标应该是评价和评估变更引起的危险与可操作性方面的风险。应评估上游和下游风险,确定安全防护措施(包括程序控制)以降低变更带来的风险。

应审查和更新技术安全分析,例如:危害辨识(HAZID)、定量风险分析(QRA)、安全完整性等级(SIL)分析、保护层分析(LOPA)、领结图等,以评估拟议的变更所带来的风险。应评估上游和下游风险并确定风险削减措施。

- **应建立变更管理(MoC)程序并有效实施**

所有变更,无论是永久的、临时的,或者紧急的,涉及程序、设施和设备、人员或物质,都应通过一个正式的管理程序来控制。

管理层已核准的变更管理(MoC)程序适用于可能影响健康安全环境(HSE)的一系列变更。变更管理(MoC)程序指明了使用哪些(子)程序管理不同类型的变更。处在安全关键岗位的变更管理(MoC)负责人和责任人,应展现其对遵守变更管理(MoC)程序的必要性的理解并积极参与。

完整的变更管理(MoC)程序包含对设备、程序、人员和组织进行永久和临时变更的所有程序。工作人员应该能展现他们知晓如何识别需要执行变更管理(MoC)程序的变更以及如何启动变更管理(MoC)程序。有效的程序应得以建立,允许人们提出改进变更管理(MoC)程序的建议。

There should be a single start point (Change Initiator or Change Request) for the identification of changes requiring MoC which should include an MoC identification table and decision tree/flow chart indicating which procedures and systems should be used for the specific MoC scope.

An MoC procedure should cover permanent, temporary and emergency changes. Emergency changes may be temporary or permanent.

The MoC procedure should include:

① Review and approval of the concept or proposal.

② Review and approval of the risk posed by the change.

③ Review and approval of any scope or design changes arising during the work.

④ Readiness review to ensure that the change has been realised. The change and risk mitigations should be built in accordance with the "issued for construction" design which has been demonstrated to be ALARP. At handover, all "as built" deviation risks for safety critical equipment should be cumulatively subject to an ALARP assessment as part of acceptance.

⑤ Requirement to inform and train the people affected by the change about what they have to do differently.

⑥ Close-out and learning capture.

A Change Control Register for temporary, permanent and emergency plant and process changes should be maintained. The Change Control Register should summarise all active change requests and the status of each request. In this way, the Change Control Register can be used to manage creeping risk of cumulative changes. The register should include an auditable trail to allow checking and close-out of updates to as-built data and documentation. Close-out dates should be stated.

应明确一个起始点(变更发起人或变更请求人)来识别需要(执行)变更管理(MoC)(程序)的变更,其中应包括变更管理(MoC)识别表和决策树/流程图,表明哪些程序和制度应该用于特定的变更管理(MoC)范围。

变更管理(MoC)程序应涵盖永久、临时和紧急变更。紧急变更可能是暂时的或永久的。

变更管理(MoC)程序应包括:

① 审查和批准设想或提案。

② 审查和批准变更带来的风险。

③ 审查和批准工作期间产生的任何范围或设计变更。

④ 审查变更就绪情况,确保变更已落实。实施的变更和风险削减措施应与满足风险可容许准则(合理可行尽可能低,ALARP)的"用于施工"的设计相一致。在交接时,安全关键设备的所有"实际建成"偏差的风险应累积在一起,作为验收的一部分按照合理可行尽可能低(ALARP)的标准进行评估。

⑤ 要求告知和培训受变更影响人员:他们的做法在变更后有何不同。

⑥ 变更关闭和经验总结。

应维护一个变更控制登记册,用于记录临时、永久和紧急的设施和工艺变更。变更控制登记册应汇总所有有效变更请求以及每个请求的状态。通过这种方式,变更控制登记册可用于管理累积变更悄然出现的风险。该登记册应包含可审核的痕迹,以允许检查和关闭对于竣工(实际建成)数据及文档的更新。应记录好每一个变更的关闭时间。

For all temporary MoCs there must be an approved expiry date, a clearly stated procedure for approval of extensions and evidence that the site demonstrates that expiry dates and deviations are audited on a monthly basis.

The Register of temporary (and temporary emergency) hardware changes should provide an auditable trail to inspection routines, temporary maintenance and function tests (i.e. integrity critical tasks). Temporary changes may override the routine systems and as such assurance is required that technical and operating integrity critical tasks are still being performed for the temporary change.

Registers of all MoC (permanent, temporary and emergency) should be in place with all relevant details logged (i.e. what is required, justification, risk and mitigating actions). The Registers should allow a full review trail from change initiation to close-out. This is demonstrable by routine sampling as part of the asset integrity assurance plan.

The risks associated with any organisational change are evaluated against the expected benefits before a decision is made to progress with change. An auditable trail is in place to demonstrate that the risks have been evaluated and any mitigating actions closed out.

- **A process should be in place to ensure that personnel involved in MoC are competent**

All personnel involved in MoC should be competent to carry out their responsibilities. They should understand the purpose, principles and practices of MoC——for their own role and for others involved in MoC. They should be aware of the site's Major Accident Hazards and the associated risks of failing to manage change to a level As Low As Reasonably Practicable (ALARP).

对所有临时变更管理(MoC)，必须有批准的到期日，有清楚规定的程序批准变更的延期，有作业场所对变更到期日和偏离进行月度审核的证据。

临时(和临时紧急)的硬件变更登记册应为例检程序、临时维修和功能测试(如，完整性关键任务)提供可审核的痕迹。临时变更可能会越过例行的程序，因此需要保证技术完整性和操作完整性关键任务仍然在为临时变更有效执行。

所有变更管理(MoC)(永久，临时和紧急)应登记在册，并记录所有相关细节(即需要的变更、理由、风险及其削减措施)。登记册应保持从变更启动到变更结束可供全面审查的痕迹。作为资产完整性保证计划的一部分，这可通过例行抽样得到证实。

在决定进行变更之前，应根据预期收益，对任何组织变更相关的风险进行评估。留下可审核的痕迹以证明风险已经评估过，任何削减措施都已到位。

- **应建立程序确保参与变更管理(MoC)的人员胜任**

所有参与变更管理(MoC)的人员在履职方面都应该是胜任的。他们应清楚变更管理(MoC)的目的、原则和做法——包括他们自己的角色以及参与变更管理(MoC)的其他人的角色。他们应该了解作业场所的重大事故危害(MAH)以及变更(风险)未能控制在合理可行尽可能低(ALARP)时的相关风险。

2.5.3 References

(1) UK HSE, HID Inspection Guide Offshore: Inspection of Safety Critical Element Management and Verification, Appendix 5, Management of changes to SCE's, Date Unknown.

(2) UK HSE, Plant Modification/ Change Procedures, http://www.hse.gov.uk/comah/sragtech/techmeasplantmod.htm, Jun, 2016.

2.5.3 参考资料

(1) UK HSE, HID Inspection Guide Offshore: Inspection of Safety Critical Element Management and Verification, Appendix 5, Management of changes to SCE's, Date Unknown.

(2) UK HSE, Plant Modification/ Change Procedures, http://www.hse.gov.uk/comah/sragtech/techmeasplantmod.htm, Jun, 2016.

3

JOURNEY MANAGEMENT AND LAND TRANSPORT SAFETY

旅程管理和道路交通安全

3.1 Journey Management

3.1.1 Prompts

(1) Have all journey hazards been identified, risks assessed and used to establish Journey Management Plans (JMPs)?

(2) Is the journey management procedure in place and operational?

(3) Do JMPs identify security as a major risk contributor and specify controls that must be in place before commencing the trip?

(4) Do JMPs consider primary and secondary routes?

(5) Are journeys actively monitored through a well-equipped Control Centre?

(6) Do reliable communication systems exist between the vehicle(s) and Control Centre during execution of the JMP?

(7) Is the journey monitored by an In Vehicle Monitoring System (IVMS)?

(8) Is there a process in place for managing periodic check-ins between the vehicle and the Control Centre?

(9) Is there a process in place to trigger an emergency response from a failed check-in?

(10) Is there a process in place to ensure that personnel who work with journey management are competent?

(11) Is a process to monitor, audit and improve the performance of the Journey Management System in place and operational?

3.1 旅程管理

3.1.1 督导提示

（1）是否已识别出旅程相关的所有危险？是否对旅程危险进行了风险评估？是否据此制定了旅程管理计划（JMP）？

（2）是否制定了旅程管理程序并有效实施？

（3）旅程管理计划（JMPs）是否将安保确定为重要风险贡献因素？旅程管理计划（JMPs）是否规定了旅程开始前必须落实的安保风险防控措施？

（4）旅程管理计划（JMP）是否考虑了首选和备选路线？

（5）是否通过设备完善的控制中心对旅程进行主动监测？

（6）在执行旅程管理计划（JMP）期间，车辆和控制中心之间是否有可靠的通信系统？

（7）是否通过车载监控系统（IVMS）监控旅程？

（8）是否建立了程序来管理车辆和控制中心之间的定期签到？

（9）是否建立了程序来启动签到失败后的应急响应？

（10）是否建立了程序来确保从事旅程管理工作的人员可以胜任（其岗位要求）？

（11）是否建立了程序来监测、审核和改进旅程管理制度的执行情况并有效实施？

3.1.2 Reading

In addition to safety risks from vehicle, driver, and road, the security and environmental risks (i.e. hot weather, heavy rain, flooding, poor visibility, etc.) that may be encountered during the journey have to be managed.

Journey management is a planned and systematic strategy to reduce land transport-related risks. It involves the planned movement of people and equipment from one place to another including communications, route, scheduled stops, hazard warnings, provisioning, break-down and other contingencies.

- **A process should be in place to identify journey specific hazards, assess risks and to establish a Journey Management Plan (JMP)**

In order to ensure journeys by road are undertaken without incident, all at risk road journeys need to be managed effectively.

Journeys by road should be avoided unless absolutely necessary. Alternatives to taking a road journey should be fully investigated. These may include alternative safer modes of transport (e.g. flying, bus, etc.), telephone calls, online meetings, video conferencing, etc.

When journey by road vehicle is unavoidable, a risk assessment should be performed which considers the journey to be taken. The journey route and other alternative routes should be identified. All potential risks along the routes should be assessed. The risk assessment should involve personnel who have in-depth experience of the journey being planned.

3.1.2 知识准备

除了来自车辆、驾驶人员、道路的安全风险之外，旅程中可能遇到的安保和环境风险(如高温天气、暴雨、洪水、低能见度等)也必须加以管控。

旅程管理是用以减少道路交通相关风险的有计划的、系统的对策(措施)。旅程管理涉及人员和设备从一个地点到另一个地点的按计划动迁，包括通信、路线、停车计划、危险警示、补给、故障和其他突发情况。

- **应建立程序识别旅程特有危险，评估其风险并制定旅程管理计划(JMP)**

为确保道路旅程平安进行，所有有风险的道路旅程都需要有效控制。

除非绝对必要，应尽量避免道路出行❶。应充分调研道路出行的备选方案，包括其他更安全的交通工具(例如飞机、公共汽车等)、电话、网络会议、视频会议等。

当道路出行不可避免时，应对旅程进行风险评估，确定旅程路线和其他备选路线。应评估沿途所有潜在风险。对计划中的旅程，有丰富经验的人员应参与其风险评估。

---

❶这里主要是指企业利用自身交通工具的道路出行。

Journey management should be risk based with proportionately more effort and controls required for non-routine journeys than for routine journeys.

Routine travel occurs within a pre-determined locale, as determined by a best practice risk assessment. Hazards are effectively addressed by established and implemented controls. Routine journeys would be expected to be no greater than that of an urban area, plant facility, or production field.

Inter-urban trips and inter-field trips should be considered to be non-routine based on an increased risk exposure due to anticipated escalating risk factors. Travel outside of a pre-determined locale requires risk assessment prior to each trip to address immediate exposures as well as factors of change and escalation. Each non-routine trip requires a JMP, requiring formal approval based on the current and anticipated risk exposure and established controls.

A Journey Management Plan (JMP) is designed to help drivers travel safely and to organise help if they get into trouble or have an accident.

- **A journey management procedure should be in place and operational**

The site should operate a journey management procedure. The journey management procedure should identify which vehicle transport modes and types of journey require formal JMPs. For each risk the measures or controls to eliminate or reduce the risk should be determined and recorded on a JMP. The JMP is a written or electronic document that communicates specific information about a trip.

旅程管理应该是基于风险的。对于非常规旅程投入的精力和控制措施应相称地高于常规旅程。

常规旅程出现在预先确定了(最佳实践)的局部区域内,(该区域)由风险评估范例确定,建立并实施的控制措施有效覆盖(常规旅程的)危险。常规旅程(行程)预期仅在城区、厂区或油区范围内。

由于预期的风险升级因素使得风险暴露增加,城际出行和跨区出行应视为非常规旅程。旅程出现在预先确定了(最佳实践)的区域外,需要每次出行前进行风险评估,既处理当前的风险也应考虑风险变化和风险升级因素。每一个非常规出行都需要制定旅程管理计划(JMP)。基于当前和预期风险暴露情况以及建立的控制措施,该计划要求获得正式批准。

旅程管理计划(JMP)旨在帮助驾驶人员安全出行,并在遇到困难或发生事故时安排帮助。

- **应建立旅程管理程序并有效实施**

工作现场应实施旅程管理程序。旅程管理程序应明确哪些车辆运输方式和哪种类型的旅程需要正式的旅程管理计划(JMPs)。每一个风险及其消除或降低措施都应予以确定并记录在旅程管理计划(JMP)中。旅程管理计划(JMP)是用于传达具体出行信息的书面或电子文件。

Prior to commencing a journey the JMP should be completed and signed off. The JMP should include the name of the driver and passengers (if any), information about the vehicle, travel route and schedule, destinations, levels of risk and contact information.

- **The journey should be monitored by a Journey Manager against the JMP for the journey**

The Journey Manager is a person who is not engaged in the journey. The Journey Manager oversees implementation of the defined journey management process, monitors progress and responds to deviations and/or emergencies.

A Journey Manager should be appointed as being responsible for monitoring the journey. The Journey Manager should perform the pre-journey briefing based on the JMP.

The JMP should interface with the check-in system. A check-in system is a communication system used to verify the well-being of the driver and passengers at pre-determined check positions along the journey. The Journey Manager should be responsible for receiving check-ins.

- **A process should be in place for managing check-ins during execution of the JMP**

The driver (or passenger) should make a call to the Journey Manager (the check-in contact) at pre-determined times and/or locations which should be specified on the JMP. The driver should pull over to a safe place at the allocated time rather than attempt to take a call whilst driving.

The Journey Manager must know from the JMP when to expect (or make) check-in calls. The Journey Manager must be available to receive check-ins (i.e. be near his phone) for the entire duration of the trip and must know what to do if the traveller does not contact them as planned. The Journey Manager must have a copy of the JMP.

应在开始旅程之前完成旅程管理计划(JMP)并签发。旅程管理计划(JMP)应包括驾驶人员和乘客(如果有)的姓名、车辆信息、出行路线和时间表、目的地、风险等级和联系信息。

- **旅程监控人员应依据旅程管理计划(JMP)对旅程进行监控**

旅程监控人员❶是没有参与旅程的人员。旅程监控人员负责监督已制定好的旅程管理流程的实施，监控旅程进度并对偏离和/或紧急情况做出响应。

应指定旅程监控人员对旅程实施监控。在旅程开始前，旅程监控人员应根据旅程管理计划(JMP)对参与旅程的人员进行旅程要点讲解。

旅程管理计划(JMP)应与签到系统衔接。签到系统是用于在旅程中的预定检查位置对驾驶员和乘客的安全状况进行核对确认的通信系统。旅程监控人员应负责接收签到信息。

- **应建立程序管理旅程管理计划(JMP)实施期间的签到**

驾驶人员(或乘客)应在规定的时间和/或地点打电话给旅程监控人员(签到联系人)。上述时间应在旅程管理计划(JMP)中予以规定。驾驶人员应在规定的时间将车辆停靠在安全的地方打电话通报情况，而不是在开车时试图打电话。

旅程监控人员必须清楚旅程管理计划(JMP)规定的电话签到时间。在整个出行期间，旅程监控人员必须确保自己能随时接听签到电话(比如，在电话附近)，并且必须清楚差旅人员如未按计划签到的备选方案。旅程监控人员必须持有一份旅程管理计划(JMP)副本(随时备查)。

---

❶企业可根据自身情况指定合适的专职或兼职"旅程监控人员"。

The check-in frequency depends on the level of risk——the greater the risk, the more frequent the check-in calls. A common default interval is every two hours.

- **Systems providing reliable communication between the Journey Manager and the driver should be established and operational**

Check-ins can be completed using a variety of means——landlines, cell phones, satellite phones, emails, text messages or two-way radios. The crucial factor in making check-ins work is that the chosen means of communication must reliably enable the worker to initiate and receive communications.

The JMP should consider communication limitations that could create gaps, e.g. poor cell service, poor Wi-Fi availability or satellite dead zones. Communication inconsistencies should be accommodated in the JMP and check-in process.

Communication protocols should always be in place for security related incidents (both inside and outside of the production facilities).

The JMP should identify multiple communication channels (radios, GSM❶ phones, satellite phones, etc.) as well as communication protocols for when and how drivers will contact base.

- **A process should be in place for when the check-in call doesn't arrive or when an emergency situation is initiated**

Most of the time, check-ins simply verify the traveller is fine and the trip is proceeding as planned. However, in the event that the check-in call does not arrive, or that it arrives but it is a passenger saying they have been involved in a crash or incident, then a procedure should be in place to initiate action.

---

❶GSM: Global System for Mobile Communication

签到频次取决于风险级别的高低——风险等级越高,签到电话的频次应越频繁。一般情况下,每2h签到一次即可。

- **应建立旅程监控人员和驾驶人员之间的可靠通信系统并有效运行**

可采用不同的方式完成签到——固定电话、手机、卫星电话、电子邮件、短信或双向无线电对讲机。使签到顺利进行的关键因素是所选择的通信方式必须使工作人员能够可靠地发出和接收消息。

旅程管理计划(JMP)应考虑可能产生(签到)中断的通信限制,例如:手机信号差、无线网络(Wi-Fi)信号差或卫星不能覆盖的区域(死区)。这些通信不稳定都应在旅程管理计划(JMP)和签到过程中加以考虑。

通信协议应始终准备就绪,以应对安保相关事件(不管在生产设施内外)。

旅程管理计划(JMP)应该确定多种通信渠道[如无线电、全球移动通信系统(GSM)电话、卫星电话等],以及规定驾驶人员何时及如何与支撑点联系的通信协议。

- **应建立未按计划签到或应急状态启动后的程序**

大多数情况下,签到系统仅仅是为了证实差旅人员一切正常,出行正在按计划进行。可是,如果没有签到,或者联系上了但一名乘客称他们遭遇了车祸或其他事故,那么应该有一个程序启动(应急响应)措施。

When an incident happens, a rapid response can reduce the severity of the incident. Emergency response should be planned ahead in order to react effectively.

- **A process to ensure that personnel who work with journey management are competent should be in place and operational**

All personnel involved in Journey Management should be competent to carry out their responsibilities. They should understand the purpose, principles and practices of journey management——for their own role and for others involved in journey management.

The Journey Manager should be competent in developing, authorising and monitoring journey risk assessments and JMP.

All drivers should be professional drivers with the required licensing in place.

3.1.3 Related Life-Saving Rules

Life-Saving Rule 4: Wear your seat belt

Life-Saving Rule 5: While driving, do not use your phone and do not exceed the speed limits

Life-Saving Rule 6: Follow prescribed journey management plan

Life-Saving Rule 15: No alcohol or drugs while working or driving

3.1.4 References

IOGP, Land transportation safety recommended practice, Report 365, Nov 2016.

如果发生事故,快速响应可以降低事故(后果)的严重程度。应该提前制定好应急响应措施以便有效应对。

- **应建立确保旅程管理人员胜任的程序并有效实施**

参与旅程管理的所有人员都应该有能力履行其职责。他们应该清楚旅程管理的目的、原则和做法——不仅清楚自己的角色也应了解其他旅程管理人员的职责。

旅程监控人员应有能力准备、批准、监督旅程风险评估和旅程管理计划(JMP)。

所有驾驶人员都应训练有素并具备相关资质。

3.1.3 相关保命法则

保命法则4:系好安全带

保命法则5:驾车时禁止使用电话,禁止超速行驶

保命法则6:遵守既定的旅程管理计划

保命法则15:严禁在工作或驾车期间饮酒或服用禁忌药品

3.1.4 参考资料

IOGP, Land transportation safety recommended practice, Report 365, Nov 2016.

3.2 Land Transport Safety

3.2.1 Prompts

(1) Is a Land Transport Safety Management System in place and operational?

(2) Are all land transport hazards identified, risks assessed and used to establish safe methods of land transport?

(3) Do all drivers know how to perform regular (pre-use and/or daily) checks of the vehicle?

(4) Are drivers trained and competent to perform the above checks?

(5) Is there a process in place to ensure that vehicles are inspected and maintained to meet their functional specifications?

(6) Does the inspection and maintenance focus on vehicle safety critical equipment and systems?

(7) Is there a process in place to ensure that vehicle loads are properly secured?

(8) Are vehicle loads checked for secureness and for overloading of the vehicle?

(9) Are vehicle capacities posted on the vehicle?

(10) Is there an In-Vehicle Monitoring System (IVMS) in place in all vehicles?

(11) Is there a process in place to ensure that all drivers and passengers wear a seat belt at all times while in a moving vehicle?

3.2 道路交通安全

3.2.1 督导提示

(1) 是否建立了道路交通安全管理制度并有效实施？

(2) 是否识别了所有道路交通危险？是否对其进行了风险评估？是否据此确立了道路交通安全(管理)方法？

(3) 所有驾驶人员是否都知道如何对车辆进行定期(使用前和/或日常)检查？

(4) 驾驶人员是否经过培训并有能力执行以上检查？

(5) 是否建立了程序确保车辆检查和维护满足车辆功能规格(要求)？

(6) 检查和维护是否关注了车辆的安全关键设备和系统？

(7) 是否建立了程序确保车载货物得到牢靠固定？

(8) 是否对车载货物的固定情况和超载情况进行了检查？

(9) 车辆上是否贴有(额定)载重能力？

(10) 是否为所有车辆配备了车载监控系统(IVMS)？

(11) 是否建立了程序确保所有驾驶人员和乘客在车辆行驶过程中始终系好安全带？

(12) Are road signs in place warning about the use of seat belts?

(13) Is it mandatory for passengers to wear seat belts?

(14) Is seat belt enforcement linked to zero tolerance disciplinary measures?

(15) Is there a process in place to ensure that distracted driving (i.e. use mobile phone while driving) and speeding is prohibited?

(16) Is site speed limit signage sufficient and appropriate for the road?

(17) Are road speed limit monitoring programmes in place?

(18) Are speed limit violations monitored and offenders notified?

(19) Is speed limit enforcement linked to zero tolerance disciplinary measures?

(20) Is there a process in place to ensure driver's competency?

(21) Are all drivers of vehicles trained, licensed and competent to drive the assigned category of vehicle?

(22) Is there a process in place to ensure driver's fitness, driving and rest hours requirements?

(23) Are drivers required to periodically undergo medical tests to confirm fitness to work?

(24) Are journeys planned to include regular breaks and restricted hours?

(25) Are drivers' working hours monitored with a restricted limit in place?

(12) 是否设立了使用安全带的道路警示标志？
(13) 佩戴安全带是对乘客的强制要求吗？
(14) 强制系安全带是否跟零容忍纪律措施关联？
(15) 是否建立程序确保分神驾驶(如：开车打手机)和超速得以禁止？
(16) 作业场所的道路限速标识是否充分并恰当？
(17) 是否编制了道路限速监测方案？
(18) 是否监测违反限速规定的行为并告知违规者？
(19) 强制限速是否跟零容忍纪律措施相关联？
(20) 确保驾驶人员能力(胜任其岗位要求)的程序是否建立？
(21) 所有车辆驾驶人员是否都经过了培训？是否获得了驾驶许可并有能力驾驶指定类别车辆？
(22) 确保驾驶人员健康适岗并按规定时间进行驾驶和休息的程序是否建立？
(23) 是否要求驾驶人员定期接受体检以确认其健康适岗性？
(24) 旅程计划是否包括了定期休息以及驾驶时间的限制？
(25) 驾驶人员的工作时长是否被监测并限定其工作时长？

(26) Is there a process in place to ensure no alcohol and no drugs at all times while in a moving vehicle?

(27) Is there a process in place to ensure the implementation of journey management plans?

(28) Are site and plant roads designed to restrict or minimise the need for reversing?

(29) Is there a process in place that ensures all vehicle manoeuvring activities keep pedestrians out of the line of fire?

(30) Is vehicle access within the plant restricted?

(31) Is a PTW required to take a vehicle into hazardous areas?

(32) Is a process to monitor, audit and improve the performance of the Land Transport Safety Management System in place and operational?

3.2.2 Reading

The effective management of land transport safety is critical to avoid loss of life and injuries (to passengers and pedestrians), property damage and reputational damage.

Land transport means transportation of goods and personnel from one place to another on road or off road. The scope of land transport safety includes:

(26) 是否建立了程序确保(驾驶人员在)车辆行驶时全程严禁酒精及违禁药品?

(27) 是否建立了程序确保旅程管理计划(JMP)得以实施?

(28) 作业场所和厂区道路设计是否限制或最小化倒车的需要?

(29) 是否建立了程序避免车辆移动时撞到行人?

(30) 是否限制车辆进入厂区?

(31) 车辆进入危险区域是否要求办理作业许可(PTW)?

(32) 是否建立并实施程序以监测、审核和提升道路交通安全管理制度的执行情况?

3.2.2 知识准备

有效管理道路交通安全,对于避免人员伤亡(包括乘客及行人)、财产损坏和声誉损害至关重要。

道路交通是指将货物和人员,沿着(正式)道路或者以越野方式,从一个地方运输到另一个地方。道路交通安全的范围包括:

① All company and contractor vehicles and drivers operating on company roads and premises.

② All company and contractor vehicles and drivers operating on public roads and in public areas on company business.

③ All transportation activities including personnel, freight and material movements, and mobile plant(drilling trucks, seismic vibrator trucks, etc.) activities on company business.

- **A Land Transport Safety Management System should be in place and used to identify land transport hazards, assess risks and to establish safe methods of land transport**

An effectively implemented management system, with due focus on land transport risk controls, can yield many benefits including improved driving safety performance and a consequential reduction in the number and severity of incidents, leading to a reduction in injuries and fatalities.

Site specific risk assessments should be performed that cover the range of land transport operations. The risk assessments should factor in local conditions, behaviours and culture.

All hazards related to land transport activities should be identified, documented and risk assessed. Where eliminating risks is not feasible, risk controls should be defined to reduce risks to an acceptable level.

Eliminating the hazard is the most effective control to reduce land transport incidents, therefore the first step in any risk assessment should be to consider whether the journey is necessary.

① 行驶在企业所属道路上和所属场所内的企业及其承包商的所有车辆和驾驶人员。

② 为执行企业事务，行驶在公共道路上和公共区域内的企业及其承包商的所有车辆和驾驶人员。

③ 所有与企业事务相关的运输活动，包括人员、货物和材料的移动以及可移动设备(如：车载钻机，物探震源车等)的活动。

- **应建立道路交通安全管理制度，用于识别道路交通危险、评估其风险并确立道路交通安全(管理)方法**

以道路交通风险防控为核心并有效实施的管理制度，可以带来许多好处，包括提高驾驶安全绩效以及减少事故数量和严重程度，进而减少人员伤亡。

覆盖不同道路交通活动的风险评估，应结合道路具体情况进行。风险评估应将当地环境、行为和文化因素考虑进来。

应识别所有与道路交通活动有关的危险，做好记录并进行风险评估。如果消除风险不可行，应制定风险防控措施将风险降低到可接受的水平。

消除危险是减少道路交通事故最为有效的控制措施。因此，任何风险评估的第一步应考虑此次差旅是否必要。

- **Only operate vehicles that are fit for purpose**

Vehicles should be fit for purpose based on an assessment of usage and should be maintained in safe working order in line with manufacturers' specifications and local legal requirements

- **Only operate vehicles with loads properly secured**

All loads should be secured to prevent damage, movement or loss during and after transit.

- **In-Vehicle Monitoring System (IVMS) should be in place and operational**

Company-owned, contracted or leased vehicles should be fitted with IVMS that records journey data for analysis.

IVMS data should be used to provide feedback to drivers and to identify drivers' performance improvement opportunities.

- **Wear a seat belt at all times while in a moving vehicle**

Drivers and all passengers should always wear a seat belt while in a moving vehicle. The Life-Saving Rule "Wear your seat belt" should be applied.

A seat belt protects you from injury in the event of an incident while driving. Wearing seat belts includes safety belts in (rental) cars, taxis, (mini) buses, trucks, cranes, or forklift trucks, and involves persons in moving vehicles.

- **Distracted driving and speeding——Do not use your phone while driving and do not exceed speed limits**

Drivers should not operate a vehicle when using a mobile phone or while being distracted from the task of driving.

Drivers should not exceed the speed limit or operate the vehicle in excess of a safe speed for the prevailing road and/or operating conditions. The Life-Saving Rule "While driving, do not use your phone and do not exceed speed limits" should be applied.

3 旅程管理和道路交通安全

- **严禁驾驶不满足要求的车辆**

应根据用途,选择合适车辆,并按照制造商提供的车辆技术规格和当地法规要求,对车辆进行维护使其处于安全工作状态。

- **严禁驾驶货物未牢靠固定的车辆**

为防止货物在运输期间和运输结束后发生损坏、位移或丢失,所有车载货物应固定牢靠。

- **应安装车载监控系统(IVMS)并确保有效运行**

企业所有、企业外包或企业租赁的车辆都应安装车载监控系统(IVMS),记录旅程数据以供分析。

应使用车载监控系统(IVMS)数据向驾驶人员提供反馈,并识别驾驶人员提升绩效的机会。

- **在行驶的车辆中,始终系好安全带**

驾驶人员和所有乘客在行驶的车辆中应始终系好安全带。保命法则"系好安全带"应得到有效实施。

车辆行驶过程中万一发生交通事故,安全带可以有效保护驾乘人员免受伤害。人员在行驶的车辆中,包括(租赁)车辆、出租车、(迷你)巴士、卡车、起重机或叉车,应系好安全带。

- **分神驾驶和超速驾驶——驾车时禁止使用电话,禁止超速行驶**

驾驶人员在使用手机时或其注意力被从驾驶任务分散时,不应驾驶车辆。

驾驶人员不得超速驾驶,或驾驶速度不应超过所在道路和/或运载状况允许的安全速度。保命法则"驾车时禁止使用电话,禁止超速行驶"应得到有效实施。

- **Driver competency**

Driver skills, knowledge and behaviours have a significant impact on driving safety.

Drivers should only operate a vehicle if appropriately trained, licensed, and competent to do so safely.

- **Driver fitness, driving and rest hours**

All persons employed as drivers and persons regularly driving on company business should undertake a driver fitness assessment to ensure that they have the functional capacity to operate a vehicle safely.

Rest hours and daily driving limits should be clearly defined.

- **No alcohol or drugs while driving**

Drivers should not operate a vehicle while under the influence of alcohol, drugs or narcotics (including illicit substances), or whilst taking medication that could impair their ability to safely operate the vehicle. The Life-Saving Rule "No alcohol or drugs while working or driving" should be applied.

- **Journey management plan**

Drivers should follow the agreed journey management plan. The Life-Saving Rule "Follow prescribed journey management plan" should be applied. A journey management plan is a plan that if followed will help drivers to travel and arrive safely.

- **Vehicle maneuvering**

Ensure pedestrians are "out of danger" when reversing a vehicle and while moving a vehicle in a work area. Working in the "line of fire" of moving vehicles is unsafe.

This practice is complementary to the Life-Saving Rule "Position yourself in a safe zone in relation to moving and energized equipment" which should be applied.

- **驾驶人员能力**

驾驶人员的技能、知识和行为对驾驶安全有重大影响。

驾驶人员仅应在参加了适当培训、获得了驾驶许可并具备安全驾驶能力时，才可以操作(驾驶许可规定类别的)车辆。

- **驾驶人员健康适岗，并遵守规定的驾驶和休息时间**

所有专职司机和因工作需要经常驾驶车辆的人员都应进行健康适岗性评估，以确保他们的身体状况满足安全驾驶要求。

(驾驶人员的)休息时间与每天驾车的(时间)限制，应有明确规定。

- **严禁在驾车期间饮酒或服用违禁药品**

驾驶人员不得在酒精、药品或毒品(包括违禁物质)影响下，或在服用可能影响其安全驾驶能力的药物时驾驶车辆。保命法则"严禁在工作或驾车期间饮酒或服用违禁药品"应得到有效实施。

- **旅程管理计划**

驾驶人员应遵守议定的旅程管理计划。保命法则"遵守既定的旅程管理计划"应得到有效实施。遵守旅程管理计划，将有助于驾驶人员安全地完成整个旅程。

- **车辆移动**

在工作区域倒车或移动车辆时，应确保不会撞到行人。在移动车辆的活动❶范围内工作是不安全的。

这个做法是对保命法则"时刻跟机动设备保持安全距离"的补充，该法则应得到有效实施。

---

❶这里的"活动"也可翻译为"危险轨迹"。

3.2.3 Related Lif-Saving Rules

Lif-Saving Rule 4: Wear your seat belt

Lif-Saving Rule 5: While driving, do not use your phone and do not exceed the speed limits

Lif-Saving Rule 6: Follow prescribed journey management plan

Lif-Saving Rule 10: Position yourself in a safe zone in relation to moving and energised equipment

Lif-Saving Rule 15: No alcohol or drugs while working or driving

3.2.4 References

(1) IOGP, Land transportation safety recommended practice, Report 365, Nov 2016.

(2) IOGP, Lif-Saving Rules, Report No. 459, April 2013, Version.

3.2.3 相关保命法则

保命法则 4：系好安全带

保命法则 5：驾车时禁止使用电话，禁止超速行驶

保命法则 6：遵守既定的旅程管理计划

保命法则 10：时刻跟机动设备保持安全距离

保命法则 15：严禁在工作或驾车期间饮酒或服用禁忌药品

3.2.4 参考资料

（1）IOGP，Land transportation safety recommended practice，Report 365，Nov 2016.

（2）IOGP，Life-Saving Rules，Report No. 459，April 2013，Version 2.

4

PERSONAL HEALTH AND PROTECTION

人员健康和保护

4.1 Food Safety

4.1.1 Prompts

(1) Is a HACCP (Hazard Analysis and Critical Control Point) Food Safety Management System in place for each food production operation?

(2) Have all food hazards been identified, risks assessed and used to establish safe methods of food production?

(3) Is a process to control food cooking in place and operational?

(4) Is a process to thaw food safely in place and operational?

(5) Is a process to cold store food safely in place and operational?

(6) Is a cleaning and sanitising programme in place and operational?

(7) Is a process for controlling the personal hygiene of food handlers in place and operational?

(8) Is a process to control pests in place and operational?

(9) Is a process in place to periodically monitor food, food contact surfaces and drinking water for microbiological pathogens?

(10) Is a process in place to handle and dispose food wastes?

(11) Is a process in place to ensure that personnel who work with food are competent?

(12) Are adequate food poisoning medical care facilities and resources in place and operational?

4.1 食品安全

4.1.1 督导提示

(1) 针对食品加工的每项操作,是否建立了基于危害分析与关键控制点(HACCP)的食品安全管理制度?

(2) 是否已识别出了所有食品危害? 是否对危害进行了风险评估并据此确立了食品加工的安全方法?

(3) 是否建立并实施了食品烹饪控制程序?

(4) 是否建立和实施了食品解冻安全程序?

(5) 是否建立和实施了食品冷藏安全程序?

(6) 是否建立和实施了清洁和消毒方案?

(7) 是否建立和实施了针对食品加工人员的个人卫生控制程序?

(8) 是否建立和实施了害虫控制程序?

(9) 针对食品、食品接触表面和饮用水,是否建立和实施了病原微生物❶定期监测程序?

(10) 是否建立和实施了厨余垃圾管理和处置程序?

(11) 是否建立和实施了确保食品加工人员胜任岗位的程序?

(12) 充足的针对食物中毒的医疗护理设施和资源是否到位并投入使用?

---

❶在本书中,Microbiological Pathogen 也翻译为致病微生物。

(13) Is the health surveillance system used to monitor foodborne illnesses in place and operational?

(14) Is a process to monitor, audit and improve the performance of the Food Safety Management System in place and operational?

4.1.2 Reading

Food safety and hygiene is critical for ensuring a safe workforce. The main risk is biological pathogens (e.g. E. Coli, Salmonella, etc.) that may lead, in the worst case, to mass food poisoning. Food poisoning is a sudden or delayed onset of illness, brought about by eating contaminated or poisonous food. The symptoms normally include abdominal pain, diarrhoea, nausea, vomiting and fever. In the worst case food poisoning can lead to fatalities if adequate medical intervention is not provided. Mass food poisoning cases can rapidly overwhelm site medical facilities requiring Medevac.

- **A HACCP Food Safety Management System should be in place in each food production operation**

Hazard Analysis and Critical Control Point (HACCP) is an internationally recognized system for reducing the risk of safety hazards in food. HACCP is also a process control system that identifies where hazards might occur in the food production process and puts into place stringent measures to prevent the hazards from occurring. HACCP includes biological (e.g. Salmonella, etc.), chemical (e.g. contamination with harmful additives, etc.) and physical hazards (e.g. contamination with insects, etc.).

（13）是否建立和实施了用于监测食源性疾病的健康监护制度？

（14）是否建立并实施程序以监测、审核和提升食品安全管理制度的执行情况？

4.1.2　知识准备

食品安全与卫生对确保员工安全至关重要。（食品安全的）主要风险来自致病微生物（如大肠杆菌、沙门氏菌等），在最坏情况下这些致病微生物可能导致大规模食物中毒。食物中毒可以是急性或延迟发病，是由进食被污染的或有毒的食物引起。临床症状通常有腹痛、腹泻、恶心、呕吐和发热。如果没有及时救治，最严重情况可导致死亡。大规模食物中毒可迅速出现现场医疗资源无法应对的情况，需要进行医疗转运。

- **应建立基于危害分析与关键控制点(HACCP)的食品安全管理制度并应用于食品加工的每项操作**

危害分析与关键控制点（HACCP）是一套旨在减少食品安全风险的国际认证体系。危害分析与关键控制点（HACCP）也是一个过程控制系统，用于发现在食品生产过程中哪个环节可能出现危害，并且采取严格的措施预防危害的发生。危害分析与关键控制点（HACCP）涵盖了生物的（如沙门氏菌等）、化学的（如有害添加剂污染等）和物理的危害（如昆虫污染）。

HACCP involves the following seven steps:

① Identify what could go wrong (the hazards).

② Identify the most important points where things can go wrong (the critical control points-CCPs).

③ Set critical limits at each CCP (e.g. cooking temperature/time).

④ Set up checks at CCPs to prevent problems occurring (monitoring).

⑤ Decide what to do if something goes wrong (corrective action).

⑥ Prove that the HACCP Plan is working (verification).

⑦ Keep records of all of the above (documentation).

By strictly monitoring and controlling each step of the process, there is less chance for hazards to occur.

Catering contractors and sub-contractors should be HACCP certified.

- **A process to control cooking should be in place and operational**

Cooked food should be stored either above 64℃ or below 5℃ to prevent food poisoning pathogens from surviving or multiplying. Cooked food that is cooled must not be at ambient temperature for longer than 90 minutes.

A digital food probe thermometer should be used to check that the core temperature of all high risk foods have achieved 75℃.

Records of cooking, cooling and hot holding temperatures should be maintained as part of the food safety management system.

Samples (100g) of all foodstuffs served during each meal should be marked and kept in a refrigerator below 5℃ for at least 72 hours after serving.

危害分析与关键控制点(HACCP)包括下列7个步骤：

① 识别可能发生的问题(危害)。

② 识别最重要的一些可能出问题的关键点(关键控制点-CCPs)。

③ 在每个关键控制点(CCP)上设置关键限值(如：烹饪的温度和时间)。

④ 在关键控制点(CCPs)上进行检查，防止问题发生(监测)。

⑤ 对发现的问题决定采取何种措施(纠正措施)。

⑥ 验证危害分析与关键控制点(HACCP)计划是否有效实施(核实)。

⑦ 将以上所有情况记录备案(文件化)。

通过对过程的每一环节进行严格监测和控制来降低危害发生的可能性。

餐饮承包商和分承包商应获得危害分析与关键控制点(HACCP)认证。

- **应建立和实施食品烹饪控制程序**

熟食应储存在高于64℃或低于5℃的环境，防止导致食物中毒的病菌生存和繁殖。冷却的熟食不应在室温下放置超过90min。

对于高风险的食物，应使用电子温度探针测量食物中心温度，确保超过75℃。

作为食品安全管理制度的一部分，应保持烹饪、冷却和加热储存温度的记录。

每餐食物均应保留样品(100g)，样品应做好标注并储存在低于5℃的冰箱至少72h。

- **A process to thaw food safely should be in place and operational**

A facility for defrosting frozen foods should be provided. This may be either a purpose built "Rapid Thaw Cabinet", a refrigerator or a chill room with a temperature below 5℃. Do not assist defrosting by placing the frozen product in water, warm oven or hot surface. Once food is thawed it should not be refrozen.

- **A process to cold store food safely should be in place and operational**

All cold storage units should have thermometers and temperatures should be monitored and recorded three times a day and records kept.

Fish and fish products should be stored in a separate freezer. Where this is not practical, fish should be placed in separate compartments or shelves. Fish, meat and poultry should be stored on shelves below fruit and vegetables to prevent contamination of the fruit and vegetables with blood.

Walk-in freezers/chillers should have slatted metal shelves and good lighting. They should be equipped with safety devices to prevent accidental lock-in. A thermometer gauge should be fixed outside the unit to give temperature readings inside the chillers/freezers. The gauge should be maintained in good working order and calibrated on a weekly basis. Records of calibration should be kept.

Ice cream and ice should not be stored in the same freezer as meat, fish or poultry to avoid cross contamination.

- **应建立和实施食品解冻安全程序**

应有专用于食品解冻的设施。它可以是快速解冻箱,也可以是温度低于5℃的冰箱和冷库。不可把食物放置在水中、暖炉或高温物体表面来帮助解冻。食物解冻后,不应再次冷冻。

- **应建立和实施食品冷藏安全程序**

所有食品冷藏单元都应有温度计,应每天3次对温度进行测量和记录,并存档。

鱼类和鱼类制品应储存在单独的冰柜中。如果不可行,鱼类应该存储在独立的(冰柜)隔间或货架上。鱼类、肉类、禽类应存放在水果蔬菜的下方,以防止其血液污染水果和蔬菜。

人可以进入的冰柜/制冷机应有覆有板条的金属货架和良好照明。它们还应配备防止冰柜/制冷机意外锁住的安全装置。温度显示应安装在冰柜/制冷机外面,以方便读取冰柜/制冷机内的温度。温度计应保持工作正常,每星期校准并加以记录。

冰激凌和冰块不应和肉类、鱼类、禽类食品冷藏在一起,以防止交叉污染。

- **A cleaning and sanitizing program should be in place and operational**

Microorganisms are one of the primary causes of spoilage and off flavours in food products. The production of consistent, high-quality food products requires the implementation of a thorough, well planned cleaning and sanitizing program aimed at controlling and/or reducing the amount of bacteria entering products during and after processing/preparation.

A specific and measurable cleaning schedule should be prepared to document what is to be cleaned, frequency of cleaning, chemicals and processes required, person responsible and personal protective equipment necessary. Supervisors should sign log sheets to confirm cleaning has taken place. A schedule should be in place and operational for each food preparation area, each food storage area and all food preparation equipment.

All fresh fruits and vegetables consumed without peeling or cooking and eaten raw are to be disinfected by washing with a food grade chlorine solution.

- **A process for controlling personal hygiene of food handlers should be in place and operational**

Unauthorised persons should not be allowed in areas where food is prepared/handled. A notice to this effect should be placed outside these areas.

- **应建立和实施清洁和消毒方案**

微生物是食品变质并产生臭味的主要原因之一。制作始终如一高品质的食品，需要实施考虑周到、安排妥善的清洁和消毒方案，以控制和/或减少食品加工/准备过程中及食品加工/准备完成后进入食品的细菌数量。

应建立具体的、可度量的时间表，用于记录清洁对象、清洁频次、所需化学品和流程、负责人员和必要的个体防护装备。主管人员应在记录表上签字，以确认清洁工作已（按清洁时间表）完成。每个食品加工区、每个食品储存区以及所有食品加工设备都应建立（清洁和消毒）时间表，并付诸实施。

所有不经削皮或烹饪直接食用的新鲜水果和蔬菜，需要用食品级的含氯溶液进行清洁消毒。

- **应建立和实施食品加工人员个人卫生控制程序**

非授权人员不能进入食品准备/处理区。在这些区域外应张贴这样的公告。

Any food handler suffering from diarrhoea, vomiting, any infectious disease, high temperature, or who has septic sores or cuts in his hands or body should immediately report to his supervisor and should be prevented from handling food until he is certified as fit to return to work by an approved Medical Practitioner.

Each food handler should be provided with a minimum number of aprons, caps and non-slip closed footwear. These must be in good repair and easy to clean.

Food handlers must have a clean and tidy appearance, clean hands with short fingernails and short hair which should be covered during food preparation. Jewellery should not be worn while working with food.

Food handlers should wear plastic disposable gloves whilst handling food. Gloves are to be replaced after each use.

Food handlers should maintain good personal hygiene. Food handlers should wash their hands on entering the food preparation area, after visiting the toilet, after coughing or sneezing, after smoking, after handling waste or carrying out cleaning activities, before touching food and between handling raw and cooked food.

- **A process to control pests should be in place and operational**

General cleanliness and good housekeeping of camps and surroundings should be maintained as the primary method of pest control. Pests of public health significance include flies, mosquitoes, cockroaches, rodents, and ants.

任何食品加工人员，如患有腹泻、呕吐、任何传染性疾病、高热、在手或身体上有感染性伤口和切口，应立即报告其主管，并停止食品加工，直到有资质的医务人员准许其返岗工作。

应为每位食品加工人员至少提供围裙、帽子和防滑封闭式的鞋子。上述物品必须无破损并易于清洁。

食品加工人员必须有干净整洁的外表、短指甲且干净的手、短头发并应在食品加工过程中用帽子将其盖住。加工食品时，不应戴首饰。

食品加工人员在加工食品时，应穿戴一次性塑料手套。每次用完后，应对手套进行更换。

食品加工人员应保持良好的个人卫生，在下列情况下应洗手：进入食品加工区前、上厕所后、咳嗽或打喷嚏后、吸烟后、处理废物或清洁活动后、接触食物之前、在处理生食和熟食之间。

- **应建立和实施害虫控制程序**

保持营地和周边环境的清洁和整洁，应作为害虫控制的首要措施。对公共健康有显著影响的害虫包括苍蝇、蚊子、蟑螂、鼠类和蚂蚁。

A pest control programme should be developed and implemented for each site.

All pesticides used must have an attached Safety Data Sheet for Chemical Products (SDS) and risk assessment outlining all control and emergency measures.

Only trained personnel shall be authorised to handle pesticides or to operate pesticide spraying equipment. Manufacturer recommended personal protective equipment (PPE) should be used.

- **A process should be in place to periodically monitor food, food contact surfaces and drinking water for microbiological pathogens**

When a food process gets out of control, the first indicator of a problem is often an increase in spoilage bacterial count. All prepared meals should be sampled and checked for spoilage bacterial count.

- **A process should be in place to handle and dispose food wastes**

Food wastes should be stored in dedicated areas which should be kept tidy, sanitised and pest proof. Waste containers should be made of impervious material and have lids.

Waste should be collected from the food premises on a daily basis and not less than twice a week from living quarters/working sites within the Camp.

每个场所都应建立和实施害虫控制方案。

使用的所有杀虫剂必须附有概述(危害)控制措施和应急措施的化学品安全技术说明书(SDS)和风险评估。

只有受过培训的人员才有权使用杀虫剂或操作杀虫剂喷洒设备,并应使用制造商推荐的个体防护装备(PPE)。

- **应建立和实施针对食品、食品接触表面和饮用水的病原微生物定期监测程序**

当食品加工过程失控,第一个出现问题的指标通常是腐败细菌数量超标。所有饭菜都应留样,并做腐败细菌数量检查。

- **应建立和实施厨余垃圾管理和处置程序**

厨余垃圾应储存在指定的整洁、卫生和防虫的区域。盛装厨余垃圾的容器应由防水材料制作,并有盖子。

应每天从营地的食品加工区收集垃圾,每周至少2次从营地的住宿区/工作区收集垃圾。

- **A process should be in place to ensure that personnel who work with food are competent**

Food Managers, Supervisors and Handlers should be trained on food hygiene.

All senior staff including senior cooks, camp boss, catering supervisors in each catering contractor company should be fully trained and certified to Intermediate Food Hygiene and Intermediate HACCP level.

There should be competent, HACCP certified Food Hygiene personnel with a dedicated role in advanced food hygiene. The Food Hygiene personnel should be competent in Advanced Food Hygiene and possess advanced HACCP certificates obtained from an internationally recognised HACCP training organisation. Training records should be maintained.

4.2 Heat Stress

4.2.1 Prompts

(1) Have all heat stress hazards been identified, risks assessed and used to establish safe methods of work?

(2) Is a Heat Stress Management Programme in place and operational during the hot summer months?

(3) Are shaded work areas provided as much as possible?

(4) Are cooled and air conditioned rest areas provided?

(5) Are fresh drinking water sources abundant and well distributed around the site?

(6) Does rehydration include provision of electrolytes?

- **应建立和实施确保食品加工人员胜任岗位的程序**

食品加工管理人员、主管人员和操作人员应接受食品卫生培训。

所有(食品加工)高级人员,包括每家餐饮承包商的高级厨师、营地经理、餐饮监督,应接受中级食品卫生和中级危害分析与关键控制点(HACCP)的全面培训并获得证书。

应由胜任且获得危害分析与关键控制点(HACCP)认证的食品卫生人员在高级食品卫生工作方面承担专门角色。食品卫生人员应胜任高级食品卫生工作,并具有国际认可的危害分析与关键控制点(HACCP)培训机构颁发的高级危害分析与关键控制点(HACCP)证书。培训记录应予以保持。

4.2 热应激

4.2.1 督导提示

(1) 是已识别出所有热应激危害?是否对其进行了风险评估?是否据此确立了安全的工作方法?

(2) 是否在炎热夏季建立和实施了热应激管理方案?

(3) 是否尽可能提供遮阳的工作区域?

(4) 是否提供凉爽和有空调的休息区?

(5) 是否在作业场所提供了足够且分布合理的新鲜饮用水?

(6) 补充水分是否包括提供含电解质的饮用水?

(7) Is drinking self-prepared salted water or eating salts prohibited?

(8) Are worker meals prepared to provide maximum nutrition and mitigation against heat stress?

(9) Are urine charts available in adequate numbers to allow the self-assessment of hydration?

(10) Does the site operate a summer working policy with respect to very high ambient temperatures?

(11) Is a process in place to ensure that personnel who work with heat stress are competent?

(12) Are adequate heat stress medical care facilities and resources in place and operational?

(13) Is a Heat Illness Protocol detailing a range of scenarios and treatment actions available?

(14) Are sufficient medical facilities and medical supplies available to respond to a credible worst case heat illness scenario?

(15) Is the health surveillance system used to monitor heat stress illnesses in place and operational?

(16) Is a process to monitor, audit and improve the performance of the Heat Stress Management Programme in place and operational?

4.2.2 Reading

Heat stress is the effect that the thermal environment has on a person's ability to maintain a normal body temperature. Physical work generates heat in the body which must be lost to the environment through sweating and evaporation. Inability to get rid of body heat adequately may result in heat illness. A hot or humid environment makes this more difficult. The need to use personal protective clothing also affects the body's ability to lose heat. Heat stress affects both mental and physical performance.

(7) 是否禁止饮用自行配置的盐水或直接吃食盐?

(8) 准备的工作餐是否提供了足够营养并减轻热应激?

(9) 是否有足够的尿液颜色图以对脱水情况进行自我判断?

(10) 作业场所是否执行有关环境超高温的夏季工作政策?

(11) 是否建立了程序确保工作中涉及热应激的人员是胜任的?

(12) 充足的热应激医疗护理设施和资源是否到位并投入使用?

(13) 热病(治疗)规范是否明确了一系列情景?对应的治疗措施阐述得是否清晰?

(14) 为响应可能最糟糕的热病情景,是否配备了充足的医疗设施和医疗用品?

(15) 用于监测热应激疾病的健康监护制度是否建立并实施?

(16) 是否建立并实施程序以监测、审核和提升热应激管理方案的绩效?

4.2.2 知识准备

热应激是热环境对人维持正常体温能力的影响。体力劳动在体内产生的热量必须通过出汗和汗液蒸发散失到环境中。人体散热能力不足可能导致热病。炎热潮湿的环境会导致这种情况加重。需要穿戴个体防护服也会影响人体的散热能力。热应激影响人的心理机能和生理机能。

There are 4 stages of heat illness: heat rash, heat cramps, heat exhaustion and heat stroke. Heat stroke is the final and most serious stage of heat illness, often resulting from exercise or heavy or prolonged work in hot environments with inadequate fluid intake. What makes heat stroke severe and life threatening is that the body's normal mechanisms of temperature control for dealing with heat stress, such as sweating, are lost. Heat stroke can be fatal and requires immediate medical attention.

- **A process should be in place to identify heat stress hazards, assess risks and establish a safe method of work**

Heat stress risk assessments should be performed for a range of climatic conditions and workplace activities. Competent HSE advisors should be responsible for performing the risk assessments which should be supported by supervisors and workers.

All outside work in hot climatic conditions should be avoided if reasonably practicable. If unavoidable, the risk assessment should consider all measures that can be implemented to reduce the risk of heat illness to a tolerable level.

Engineering control measures should be identified to eliminate the risk of heat exposure as far as reasonably practicable. Engineering controls include adding insulation to ceilings to minimise solar heat transfer; providing shaded work areas as much as possible; providing cooled and air conditioned rest areas with water or electrolyte drinks available (not salt tablets or salt water); using exhaust ventilation such as extraction hoods above heat-generating processes; using forced air-ventilation such as fans to increase airflow across the skin and increase evaporation and cooling; and using cooled air from an air-conditioning system.

热病有 4 个阶段：热皮疹、热痉挛、热衰竭和中暑。中暑是热病的最终和最严重阶段，常见于在炎热环境中运动或高强度、长时间的工作，同时没有及时补充足够水分。中暑后果严重和威胁生命的原因是在热应激情况下，人体应对热应激的正常体温控制机制失灵，如无法正常排汗。中暑可以导致生命危险，需要立即进行救治。

- **应建立程序识别热应激危害，评估其风险并据此确立安全的工作方法**

应对一系列气候条件和工作场所活动开展热应激风险评估。胜任的 HSE 顾问应负责在主管人员和（其他）工作人员的支持下开展风险评估。

如果合理可行的话，应避免在炎热天气条件下的所有户外作业。如果无法避免，风险评估应考虑实施所有可以实施的措施将热病风险降低到可容许水平。

只要合理可行，应尽可能识别出工程控制措施来消除热暴露的风险。工程控制措施包括增加屋顶隔热层来降低日光热量的传导；尽可能多地提供遮阳的工作环境；提供凉爽和有空调的休息区，并提供饮用水或电介质饮料（不能用食盐片剂或盐水）；使用排气通风装置，如产生热量的工艺过程上方的抽气罩；使用强力通风装置，如风扇，增强皮肤上的空气流动进而增强汗液蒸发和降温效果；使用空调系统产生的冷空气。

Fresh drinking water sources should be abundant and well distributed around the site. Rehydration should include provision of electrolytes. Drinking self-prepared salted water or eating salt should be avoided by all workers.

- **A Heat Stress Management Programme (HSMP) should be in place and operational during the hot season**

The HSMP should consider all aspects of heat stress management and should be especially based on arid environment climatic conditions which are considered to be some of the severest working conditions in the world.

- **A process should be in place to ensure that personnel who work with heat stress are competent**

Training for exposed workers should include: the hazards of working in heat; the importance of maintaining good hydration (drinking at least 2 litres of water every 2-3 hours); eating a well-balanced diet and adding a little extra salt to their meals; recognizing the signs of heat illness; the hazards of consuming alcohol, tea, coffee and caffeinated drinks which may increase fluid loss; explanation of the self-assessment of hydration using the urine charts; the importance of rest and recovery and getting a good night's sleep; first aid measures to apply in case of heat illness and familiarisation with the Medical Emergency Response Plan (MERP) of every working area.

- **Heat stress medical care facilities should be in place and operational**

Workers who suffer a suspected case of heat illness should have immediate access to medical care.

Medical facilities at site should be well prepared for a range of heat illnesses, including heat stroke.

在工作场所应有足够的和分布合理的新鲜饮用水。补充水分应包括提供含电解质的饮用水。所有员工应避免饮用自行配置的盐水或直接吃食盐。

- **应在炎热季节建立和实施热应激管理方案（HSMP）**

热应激管理方案（HSMP）应考虑热应激管理的所有方面，特别是应考虑干旱地区的环境气候情况，因为它们被认为是世界上工作条件最恶劣的地区之一。

- **应建立程序确保工作中涉及热应激的人员胜任**

热环境下工作人员的培训应包括：热环境下工作的危害；身体保持充足水分的重要性（每2~3h至少饮用2L水）；膳食平衡并且在食物中稍微多加一些盐；能够识别热病的症状；了解饮用酒、茶、咖啡和含咖啡因饮料的副作用，它们能够加速水分流失；使用尿液颜色图来自我判断脱水情况；休息和恢复以及夜间良好睡眠的重要性；了解应对热病的急救措施并且熟悉每个工作区域的医疗应急响应计划（MERP）。

- **应建立并运行热应激医疗救护点**

怀疑患有热病的员工应立即送往医疗救护点。

现场医疗点应能救治不同的热病，包括中暑。

A Heat Illness Protocol should provide examples of a range of scenarios and treatment actions.

There should be sufficient medical facilities and medical supplies available to respond to a credible worst case heat illness scenario.

4.2.3 References

(1) Health Authority Abu Dhabi (HAAD), 2010—2018, Safety in the Heat, https://www.haad.ae/safety-in-heat/.

(2) OSHA, Heat Stress, 2018, https://www.osha.gov/SLTC/heatstress/.

4.3 Noise and Hearing Conservation

4.3.1 Prompts

(1) Have all noise hazards been identified, risks assessed and used to establish safe methods of work?

(2) Have all noise sources been identified with noise emission levels correctly characterised?

(3) Are full noise surveys performed on a periodic basis, not less than every two years?

(4) Are spot noise measurements using a hand held meter routinely made and focused on high noise areas?

(5) Are engineering noise controls such as baffles, enclosures, shields, lagging, dampers, etc. in place, functional and well maintained?

热病(治疗)规范应明确一系列情景并清晰阐述对应的治疗措施。

为响应可能最糟糕的热病情景，应配备充足的医疗设施和医疗用品。

4.2.3 参考资料

（1）Health Authority Abu Dhabi（HAAD），2010—2018，Safety in the Heat，https：//www.haad.ae/safety-in-heat/.

（2）OSHA，Heat Stress，2018，https：//www.osha.gov/SLTC/heatstress/.

4.3 噪声与听力保护

4.3.1 督导提示

（1）是否已识别出所有噪声危害？是否对其进行了风险评估？是否据此确立了安全的工作方法？

（2）是否已识别出所有噪声源并正确描述了其噪声水平？

（3）是否定期开展全面的噪声调查且调查频次不少于2年一次？

（4）是否使用手持仪器通过点噪声测量法经常性地对高噪声区域的噪声水平进行测量？

（5）是否使用了隔板、围挡、护罩、衬套、减振器等噪声控制的工程措施？相关工程控制措施是否功能正常并得到了良好维护？

(6) Are installations and equipment identified as being sources of significant noise and vibration included in the inspection and maintenance programme?

(7) Is a Hearing Conservation Programme (HCP) in place and operational?

(8) Are different types of hearing protection provided to workers (and visitors) based upon the magnitude and duration of the noise exposure?

(9) Is the need for, the limitations of, and the correct use of hearing protectors clearly explained to the effected workers?

(10) Is hearing protection periodically tested, repaired and maintained?

(11) Is a process in place to ensure that personnel who work in the HCP are competent?

(12) Is there a health surveillance system in place to monitor potential hearing loss in workers that are exposed to noise?

(13) Are noise exposed workers required to periodically undergo medical tests to confirm fitness to work?

(14) Does the site perform audiometric tests for noise exposed or vulnerable workers?

(15) Is a process to monitor, audit and improve the performance of the HCP in place and operational?

4.3.2 Reading

Noise is generated during processes, operations and work activities. It is one of the most common occupational health hazards. Noise-induced hearing loss can be temporary or permanent. Temporary hearing loss results from short-term exposures to noise, with normal hearing returning after a period of rest. Generally, prolonged exposure to high noise levels over a period of time gradually causes permanent hearing loss.

(6) 检维修计划是否包含了已识别出的作为重大噪声源和振动源的装置及设备？

(7) 是否建立了听力保护计划(HCP)并有效实施？

(8) 是否基于噪声量级和暴露时长为员工(和访客)提供不同类型的听力保护用品？

(9) 是否向暴露于噪声环境中的员工清楚说明了使用听力保护用品的必要性、局限性以及其正确使用方法？

(10) 是否对听力保护措施进行定期测试、维修和维护？

(11) 是否建立了程序以确保听力保护计划(HCP)所涉及的人员是胜任的？

(12) 是否建立了健康监护制度监测暴露于噪声环境中的员工的潜在听力损失？

(13) 是否要求暴露于噪声环境中的员工定期体检以确认其是否健康适岗？

(14) 作业现场是否为暴露于噪声环境或易受噪声伤害的员工进行了听力测试？

(15) 是否建立并实施程序以监测、审核和提升听力保护计划(HCP)的执行情况？

4.3.2 知识准备

噪声产生于工艺过程、操作运行和作业活动中。它是最常见的职业健康危害之一。噪声诱发的听力损失可能是暂时性，也可能是永久性的。暂时性听力损失是由于短时间暴露于噪声环境中所导致的，经过一段时间休息后听力可恢复正常。一般来说，长时间暴露于高噪声环境中，可逐渐引发永久性听力损失。

Prolonged exposure to excessive noise can cause permanent noise induced hearing loss, commonly known as noise-induced deafness (NID). NID leads to communication difficulties, impairment of personal relationships, social isolation and poor quality of life.

- **A process should be in place to identify noise hazards, assess risks and to establish a safe method of work**

A risk assessment must be conducted and documented for all noisy activities, including works, operations and processes. The first step of risk assessment involves hazard identification which can be qualitative or quantitative.

Qualitative hazard identification can be performed using a checklist or through site inspection. If the outcome of the site inspection suggests that noise problems exist in the workplace, the hazard should be quantified and evaluated through noise surveys and noise mapping of the workplace.

Risk assessments should be reviewed and revised at least three yearly, upon occurrence of NID or significant change in work practices or procedures.

Noise control measures should be used as the first priority to minimise the risk of noise exposure. If elimination of the noise source is not possible, engineering controls should be implemented. Noise can be reduced through vibration controls, including for example damping or lagging of vibrating surfaces, proper balancing and maintenance of machinery. Mufflers or silencers can control noise generated by turbulent air flow.

长时间暴露于过大噪声，可导致永久的噪声性听力损失，通常被称为噪声性耳聋(NID)。噪声性耳聋(NID)导致沟通困难、人际关系障碍、社交孤立和生活质量变差。

- **应建立程序识别噪声危害，评估其风险并确立安全的工作方法**

应对所有产生噪声的活动进行风险评估并记录存档，包括作业活动、操作运行和工艺过程。风险评估的第一步是危害辨识，可以通过定性或定量的方法。

定性危害辨识通过使用检查表或现场检查来执行。如果现场检查的结果认为工作场所存在噪声问题，就需要通过工作场所噪声调查和噪声定位来对噪声危害进行量化和评价。

根据噪声性耳聋(NID)出现的情况或者是作业方法或程序的重大变化情况，每年应至少3次对(噪声)风险进行重新评估和修订。

应优先采用噪声控制措施来最小化噪声暴露的风险。如果无法消除噪声源，就应实施工程技术措施。可以通过控制振动来减小噪声，比如：对振动物体表面进行减振处理或增加隔音材料，恰当地平衡和维护机器。消声器或消音器可以控制气体湍流产生的噪声。

Damping is the process of reducing vibration of a surface thereby reducing the level of the noise caused by the vibration. This can be achieved by using absorbing materials or fluids. Lagging is a flexible, sound absorbing material that is used to dampen noise caused by vibration. It is often used to cover pipes and ducts to block the noise that is generated when air or fluids flow through them.

Noise control can be achieved by complete or partial enclosure of the noise source by placing an acoustical shield or barrier wall between the source and the receiver, or by increasing the distance between them. The installation of acoustical absorbing materials on ceilings or walls may result in significant reduction of noise.

Hearing protectors serve to protect the employees against excessive noise during the interim period before the noise is successfully reduced through engineering control measures, or when engineering or administrative measures are not feasible. Before hearing protectors are issued to the affected employees, the need for and limitations of their use should be fully explained.

- **A Hearing Conversation Programme (HCP) should be in place and operational**

An HCP is designed to protect workers with significant occupational noise exposures from hearing impairment even if they are subject to such noise exposures over their entire working lifetimes. An HCP is required when any person in the workplace is exposed to excessive noise, which is defined as an equivalent sound pressure level of 85 dB(A) or more over an eight-hour workday.

An effective HCP can eliminate or minimise the noise hazard and prevent NID. Strong commitment by the management and active involvement by the employees are critical for the success of the HCP. Management should take the appropriate steps to encourage employees' participation in the development and implementation of the HCP.

减振是一种通过减少物体表面振动来降低振动类噪声级别的过程。这可以通过使用吸收振动的材料或液体来实现。隔音材料是一种有弹性的吸音材料，用于抑制振动产生的噪声。它常用于包裹管道或导管，阻挡因空气或液体流过管道或导管而产生的噪声。

噪声控制可以通过在噪声源和接收者之间设置隔音罩或隔音屏(墙)，将噪声源全部或部分封闭，或者通过增加噪声源和接收者的距离来实现。天花板或墙壁安装吸音材料，可以显著减低噪声。

在通过工程技术措施成功减少噪声之前，或者当工程技术或行政管理措施不可行时，可以暂时使用听力保护用品保护员工免受过大噪声的损伤。在将听力保护用品发放给噪声环境中工作的员工之前，需要给他们详细说明使用听力保护用品的必要性和局限性。

- **应制定听力保护计划(HCP)并有效实施**

听力保护计划(HCP)旨在保护员工免于因暴露于显著职业噪声环境而导致听力损伤，即使他们在整个工作生涯都暴露在噪声环境中。暴露于工作场所过大噪声环境中的任何员工，需要实施听力保护计划(HCP)。这里，过大噪声指的是在一个 8h 工作日中，等效声压级等于或大于 85 dB(A)。

有效的听力保护计划(HCP)可以消除或减小噪声危害，预防噪声性耳聋(NID)。管理层强有力的承诺和全员的积极参与是成功实施听力保护计划的关键。管理层应采取适当的方法鼓励员工参与到听力保护计划(HCP)的开发和实施中。

- **A process should be in place to ensure that personnel who work in the HCP are competent**

Educational programmes should be provided to all personnel involved in the HCP in order to raise their awareness and understanding of the noise hazard and prevention of NID.

Training is necessary for employees to understand the importance of protecting their hearing. They will then be more motivated to actively participate in the programme. Besides noise-exposed employees, supervisors and managers who are responsible for the noisy areas, and other persons involved in the HCP should be included in the training. The training should increase employees' awareness of the noise hazard and approaches they can adopt to take care of their hearing.

- **Medical surveillance process should be in place and operational as part of the HCP**

Annual audiometric examinations should be done to detect early hearing impairment and can be used to monitor the effectiveness of the HCP in preventing NID.

Audiometric examinations are an important part of the HCP as employees with mild hearing loss can be identified early. The symptoms of NID do not manifest until a significant threshold shift has occurred and workers may therefore suffer from significant NID without being aware of it. Such early detection of hearing loss will provide the opportunity for immediate measures to be taken to prevent further deterioration. The results of the audiometric tests can give an indication if the HCP is working effectively. Hearing loss can be due to:

- **应建立程序确保听力保护计划(HCP)所涉及的人员胜任**

应向听力保护计划(HCP)涉及的所有工作人员提供相关教育培训,来增强他们对噪声危害和预防噪声性耳聋(NID)的认识和理解。

开展培训让员工了解保护听力的重要性是非常必要的。这样,员工才能更加积极主动地参与到听力保护计划中来。除了暴露于噪声中的员工,对负责噪声区域的主管及负责人,还有其他跟听力保护计划(HCP)有关的人员都应进行培训。培训应增强员工对噪声危害以及他们可以采取的听力保护方法的意识。

- **应建立医学监护程序并作为听力保护计划(HCP)的一部分来实施**

应开展年度听力检查以筛查早期听力损伤,它同时也可用于监测听力保护计划(HCP)在预防噪声性耳聋(NID)方面的效果。

听力检查是听力保护计划(HCP)的一个重要部分,这样,员工听力的轻度损伤可被更早地筛查出来。只有明显的听力阈值发生变化后,噪声性耳聋(NID)的症状才表现出来,因此员工有可能患有严重的噪声性耳聋(NID)而自己却浑然不知。这种听力损失的早期筛查为立即采取措施预防病情恶化提供了机会。听力检查结果可以作为听力保护计划(HCP)是否有效实施的一项指标。导致听力损失的原因有:

① Inadequate engineering and/or administrative controls.

② Workers not using their hearing protection.

③ Changes in the work environment e. g. new machines with higher noise levels or existing machines becoming noisier due to wear and tear of bearings, machine mountings.

4.3.3 References

(1) UK HSE, Controlling Noise at Work, The Control of Noise at Work Regulations 2005, Guidance on Regulations, L108, (Second edition), Published 2005.

(2) OSHA Technical Manual, Section III: Chapter 5, Noise, August 2013.

4.4 Personal Protective Equipment

4.4.1 Prompts

(1) Have all workplace hazards been identified, risks assessed and used to establish safe methods of work and PPE requirements?

(2) Is a PPE procedure in place and operational?

(3) Is required PPE provided, maintained and used?

(4) Does all PPE meet functional, reliability and operating requirements?

(5) Is a PPE matrix (stating task and PPE) available for the specific area and activity?

(6) Is suitable storage for PPE available and adequate?

(7) Are protective helmets used wherever there is a possible danger of head injury from impact, or from flying objects, or from electrical shock and burns?

① 不适当的工程技术和/或行政管理措施。
② 工作人员没有采用听力保护措施。
③ 工作环境的变化。例如：(使用了)高噪声的新机器或现有机器因为轴承和机器固定件的磨损导致噪声增加。

4.3.3 参考资料

(1) UK HSE, Controlling Noise at Work, The Control of Noise at Work Regulations 2005, Guidance on Regulations, L108, (Second edition), Published 2005.

(2) OSHA Technical Manual, Section Ⅲ: Chapter 5, Noise, August 2013.

4.4 个体防护装备

4.4.1 督导提示

(1) 是否已识别出工作场所的所有危害？是否对危害进行了风险评估？是否据此确立了安全的工作方法以及个体防护装备(PPE)的要求？

(2) 是否建立了个体防护装备(PPE)程序并有效实施？

(3) 是否按要求提供、维护并使用个体防护装备(PPE)？

(4) 所有个体防护装备(PPE)是否满足功能、可靠性及作业要求？

(5) 是否为特定区域和特定作业活动制定了个体防护装备(PPE)矩阵(建立作业任务和个体防护装备之间的对应关系)？

(6) 个体防护装备(PPE)储备是否(类型)恰当且(数量)充足？

(7) 当存在可能造成头部伤害的危险时，如冲击(撞击、碰撞)、飞行的物体、电击和电灼伤等，是否使用了防护头盔？

(8) Are appropriate protective gloves used wherever there is the danger to hands of exposure to hazards such as those from skin absorption of harmful substances, severe cuts or lacerations, severe abrasions, punctures, chemical burns and thermal burns?

(9) Is appropriate protective footwear used wherever there is the danger of foot injuries due to heat, corrosive and poisonous substances, falling or rolling objects, or objects piercing the sole, and where feet are exposed to electrical hazards?

(10) Are individuals issued and required to wear appropriate eye protective devices while participating or observing activities which present a potential eye safety hazard?

(11) Are Self Contained Breathing Apparatus (SCBA) adequately provided?

(12) Are noise protection devices being used and used correctly?

(13) Is appropriate PPE for electrical protection available where required?

(14) Is fire retardant clothing available?

(15) Is a process in place to ensure that PPE is well stocked and available at all times?

(16) Is a process to control the inspection and maintenance of PPE in place and operational?

(17) Are hard hats inspected periodically for damage to the shell and suspension system?

(18) Are respirators inspected, maintained regularly and are they stored correctly?

(19) Are personnel trained in the use of SCBA?

(20) Is a process in place to ensure that personnel that work with PPE are competent in all aspects of PPE usage?

（8）在因暴露于某些危害，如皮肤吸收有害物质、严重切割或撕裂伤、严重擦伤、穿刺、化学灼伤和热灼伤等，而存在手部危险的地方是否使用合适的防护手套？

（9）在因热、腐蚀性和有毒物质、坠落或滚动物体、刺穿鞋底的物体以及脚部暴露于电气危害而存在脚部受伤危险的地方，是否使用合适的防护鞋？

（10）当参与或观察的作业活动对眼部可能带来危害时，是否给参与者或观察者分发并要求其佩戴适当的眼部防护用品？

（11）是否配备了充足的正压呼吸器（SCBA）？

（12）是否正在使用噪声防护用品并使用正确？

（13）当需要时，是否可得到合适的用于电气防护的个体防护装备（PPE）？

（14）（当需要时）是否可得到阻燃服？

（15）是否建立了程序确保个体防护装备（PPE）库存充足并随时可用？

（16）是否建立了个体防护装备（PPE）检查和维护的控制程序并有效实施？

（17）是否定期检查安全帽壳体和悬挂系统的损坏情况？

（18）是否对呼吸器进行定期检查、保养？它们的存放是否正确？

（19）所有使用正压呼吸器（SCBA）的人员是否都进行了培训？

（20）是否建立了程序确保个体防护装备（PPE）使用者在个体防护装备（PPE）使用的各个方面都是胜任的？

4.4.2 Reading

Personal Protective Equipment (PPE) is equipment that will protect the user against health or safety risks at work. It can includes, but is not limited to, items such as safety helmets, gloves, eye protection, high-visibility clothing, safety footwear, safety harnesses, etc.

PPE should be used as a last resort. Wherever there are risks to health and safety that cannot be adequately controlled in other ways, then PPE must be used.

• **A process should be in place to identify workplace hazards, assess risks and identify PPE requirements to establish a safe method of work**

To make sure the right type of PPE is chosen, the different hazards in the workplace must be risk assessed with PPE identified that will provide adequate protection against them.

The following should be considered when assessing the suitability of PPE:

① PPE should protect the wearer from the risks and should be able to cater for the prevailing environmental conditions where the task is taking place.

② PPE should not increase the overall level of risk or add new risks, e.g. by making communication more difficult.

③ PPE should be adjustable to fit the wearer correctly.

④ PPE selection should be sensitive to the needs of the job and the ergonomic and physical demands it places on the wearer (it should not make the job significantly more difficult and should not be difficult or very uncomfortable to wear).

4.4.2 知识准备

个体防护装备(PPE)是在工作中保护使用者免受健康或安全风险的装备(或用品)。个体防护装备(PPE)可包括但不限于诸如安全帽、手套、眼部防护用品、高能见度的服装、安全鞋、安全带等物品。

个体防护装备(PPE)应作为最后的安全措施使用。在其他控制措施都不能充分地控制健康与安全风险时,那么就必须使用个体防护装备(PPE)。

- **应建立程序识别工作场所的危害,评估其风险并确定个体防护装备(PPE)要求以确立安全的工作方法**

为确保正确选择个体防护装备(PPE)类型,工作场所不同危害的风险评估必须结合个人防护装备(PPE)的确定,从而确保确定的个体防护装备(PPE)将会提供足够的风险防护。

评估个体防护装备(PPE)的适用性时应考虑以下内容:

① 个体防护装备(PPE)应为佩戴者提供风险防护,并应能够满足目前所执行任务的主导环境条件。

② 个体防护装备(PPE)不应增大总的风险等级或加入新的风险,例如:使交流更加困难。

③ 个体防护装备(PPE)应可调整,以适合不同佩戴者。

④ 个体防护装备(PPE)的选择应该特别考虑工作需要和人体工程学及穿戴者身体的要求(它不应使工作明显变得更困难,不应穿戴困难或穿戴非常不舒服)。

⑤ PPE should be compatible with other PPE that has been selected. For example, does using a respirator make it difficult to fit eye protection properly?

PPE should be used to provide protection to any part of the body that is at risk such as the head (helmet), the eyes (goggles), the ears (ear muffs), the lungs (respirator), the feet (steel toeboots) and the hands (gloves).

- **A PPE procedure should be in place and operational**

A procedure should be established to control PPE. Where PPE is used to perform work as part of the PTW system then each item of PPE used should be determined from the JHA. For routine work activities performed outside of the PTW then the process used to define and justify PPE selection should be provided in the best practice JHA.

- **A process to control the inspection and maintenance of PPE should be in place and operational**

The process should ensure that:

① PPE is well looked after and properly stored when it is not being used, e.g. in a dry, clean cupboard, or for smaller items in a box or case.

② PPE is kept clean and in good repair follow the manufacturer's maintenance schedule (including recommended replacement periods and shelf lives).

③ Simple maintenance can be carried out by the trained wearer, but more intricate repairs should only be done by specialists.

⑤ 一种个体防护装备(PPE)应与已选择的其他个体防护装备(PPE)兼容。例如：使用呼吸器是否会导致正确使用护目镜变得困难？

个体防护装备(PPE)应用于保护身体任何处在风险中的部位，如头部(头盔)、眼部(护目镜)、听力(耳罩)、肺部(呼吸器)、脚部(钢头靴子)和手部(手套)。

- **应建立个体防护装备(PPE)程序并有效实施**

应建立个体防护装备(PPE)控制程序。对于需要作业许可(PTW)系统来控制的作业任务，当需要使用个体防护装备(PPE)时，所使用的每件个体防护装备(PPE)都应通过"工作危害分析(JHA)"来确定。对于不需要作业许可(PTW)系统来控制的日常作业活动，确定和论证个体防护装备(PPE)的过程应在工作危害分析(JHA)范例中提供。

- **应建立个体防护装备(PPE)检查和维护的控制程序并有效实施**

该程序应确保：

① 个体防护装备(PPE)在不用时应妥善照看并正确存放，例如：存放在干燥、洁净的橱柜中，或将小物件存放在盒子或箱子里。

② 应按制造商的维护计划(包括建议的更换期限和保存期限)，保持个体防护装备(PPE)清洁并保持良好维护。

③ 简单的维修可以由受过训练的佩戴者进行，但复杂的维修只应由专业人员进行。

④ Replacement parts match the original, e.g. respirator filters.

⑤ Employees make proper use of PPE and report its loss, destruction and any faults.

- **A process should be in place to ensure that personnel who work with PPE are competent**

Everyone using PPE should be aware of why it is needed, when to use, repair or replace it and how to report it if there is a fault or limitation.

① Train and instruct people how to use PPE properly and make sure they are doing this. Include managers and supervisors in the training, they may not need to use the equipment personally, but they do need to ensure their staff are using it correctly.

② It is important that users wear PPE all the time they are exposed to the risk. Never allow exemptions for those jobs which take "just a few minutes".

③ Check regularly that PPE is being used and investigate incidents where it is not.

④ Safety signs can be useful reminders to wear PPE, make sure that staff understand these signs, what they mean and where they can get equipment.

4.4.3 References

UK HSE, Personal protective equipment (PPE) at work. A brief guide. INDG174(rev2), Published 06/13.

④ 更换的部件应与原件相匹配，如：呼吸器的过滤器。

⑤ 员工应正确使用个体防护装备(PPE)，并报告其丢失、损坏以及任何故障。

- 应建立程序确保佩戴个体防护装备(PPE)的人员胜任(装备的使用)

使用个体防护装备(PPE)的每个体都应该知道为什么需要使用个体防护装备(PPE)？何时使用、修理或更换？个体防护装备(PPE)有缺陷或局限时，如何报告？

① 培训和指导员工如何正确使用个体防护装备(PPE)，并确保他们付诸于行动。管理人员和主管人员可能不需要亲自使用个体防护装备，但应将他们纳入培训范围，因为他们需要确保他们的下属正确使用个体防护装备(PPE)。

② 一旦暴露于危险之中，必须一直佩戴个体防护装备(PPE)。在这种情况下，坚决杜绝对"几分钟不佩戴PPE"的豁免。

③ 定期检查个体防护装备(PPE)的使用情况，并调查不佩戴个体防护装备(PPE)的事件。

④ 可以通过安全标识提醒员工佩戴个体防护装备(PPE)，确保员工明白这些标识，包括标识的含义以及在哪里可以获取防护装备。

4.4.3 参考资料

UK HSE, Personal protective equipment (PPE) at work. A brief guide. INDG174(rev2), Published 06/13.

APPENDIX

附录

Appendix 1 Leadership Site Visit Prompt Card Sample

LEADER NAME <Insert Leader Name >		
SITE NAME <Insert Location>		
DATE <Insert Date >		

LSV PROMPT CARD
<INSERT TOPIC>

LSV-PC-XX

PROMPT <Insert Prompt>	SCORE
Supporting Evidence?	
PROMPT <Insert Prompt>	SCORE
Supporting Evidence?	
PROMPT <Insert Prompt>	SCORE
Supporting Evidence?	
PROMPT <Insert Prompt>	SCORE
Supporting Evidence?	
PROMPT <Insert Prompt>	SCORE
Supporting Evidence?	
PROMPT <Insert Prompt>	SCORE
Supporting Evidence?	
PROMPT <Insert Prompt>	SCORE
Supporting Evidence?	
PROMPT <Insert Prompt>	SCORE
Supporting Evidence?	

Remark: For how to use LSV PROMPT CARD, please refer to relevant sections of the Book—Recommended Leadership Site Visit Program and Safety Models.

附录1 管理人员现场安全督导提示卡样例

管理人员姓名 <填写管理人员姓名>		**督导提示卡**	
地点 <填写地点>		**<嵌入主题>**	
时间 <填写年月日>			LSV-PC-XX

	打分		打分
督导提示 <嵌入督导提示>	☐	督导提示 <嵌入督导提示>	☐
支持性证据?		支持性证据?	
督导提示 <嵌入督导提示>	☐	督导提示 <嵌入督导提示>	☐
支持性证据?		支持性证据?	
督导提示 <嵌入督导提示>	☐	督导提示 <嵌入督导提示>	☐
支持性证据?		支持性证据?	
督导提示 <嵌入督导提示>	☐	督导提示 <嵌入督导提示>	☐
支持性证据?		支持性证据?	

注：管理人员现场安全督导提示卡的使用说明参见分册《安全督导推荐做法及常用安全模型》的有关章节。

APPENDIX

LSV PROMPT CARD
<INSERT TOPIC>

LSV-PC-XX

SITE NAME
<Insert Location>

SAFETY CONVERSATION

As part of the site visit you will have safety conversations with the time organisation.
Use this form to summarise your observations.

- Who did you talk to? (Provide position not name)
- How was your safety conversation received?
- How well did they understand the topic?
- Were they able to show you procedures they use to control their work?
- How well do they understand and implement the procedures?
- What information/suggestions did you provide to them?
- What improvements did they suggest?

NOTES

LEADER NAME	POSITION
SIGNATURE	DATE

附录

督导提示卡
<嵌入主题>

LSV-PC-XX

地点
<填写地点>

安全交谈
作为现场安全督导工作的一部分，您将与现场人员进行安全交谈。请使用该检查表记录您的安全谈所得。

您与谁进行了交谈？(提供岗位而不是姓名)

你们是如何开展安全交谈的？

他们对本次交谈的主题理解得怎么样？

他们能否向您展示用于控制作业（危害）的程序？

他们对流程理解得怎么样？实施得怎么样？

您向他们提供了什么信息/建议？

他们提出了什么改进建议？

备注

管理人员姓名	岗位
签名	日期

Appendix 2 Abbreviations and Acronyms

附录 2 缩略语

AIChE	American Institute of Chemical Engineers
ALARP	As Low as Reasonably Practicable
CCP	Critical Control Point
EBM	Evidence-Based Medicine
ECC	Emergency Control Centre
ERC	Emergency Response Centre
ERP	Emergency Response Plan
GSM	Global System for Mobile Communications
HACCP	Hazard Analysis and Critical Control Point
HAZID	Hazard Identification
HAZOP	Hazard and Operability
HCP	Hearing Conservation Programme
HRA	Health Risk Assessment
HSE	Health, Safety and Environment
HSECES	HSE Critical Equipment and Systems
HSE-MS	Health, Safety and Environment Management System
HSMP	Heat Stress Management Programme
IADC	International Association of Drilling Contractors
ICS	Incident Command System
IMT	Incident Management Team

IOGP	International Association of Oil & Gas Producers
IVMS	In Vehicle Monitoring System
JHA	Job Hazard Analysis
JMP	Journey Management Plan
LOPA	Layer of Protection Analysis
LSVP	Leadership Site Visit Program
MAH	Major Accident Hazard
MERP	Medical Emergency Response Plan
MoC	Management of Change
MOPO	Manual of Permitted Operations
NID	Noise-Induced Deafness
OSHA	Occupational Safety and Health Administration
PPE	Personal Protective Equipment
PTW	Permit to Work
QRA	Quantitative Risk Assessment
SCBA	Self-Contained Breathing Apparatus
SDS	Safety Data Sheet for Chemical Products
SERP	Site Emergency Response Plan
SIL	Safety Integrity Level
SIMOPS	Simultaneous Operations
SME	Subject Matter Expert
SOP	Standard Operating Procedures
TBT	Toolbox Talks
UK HSE	UK Health and Safety Executive
WHO	World Health Organization
WI	Work Instructions

Appendix 3　English-Chinese Vocabulary

附录3　中英文对照词汇表

Adverse Weather	恶劣天气
American Institute of Chemical Engineers(AIChE)	美国化学工程师协会
As Low as Reasonably Practicable (ALARP)	合理可行尽可能低
Asset Integrity	资产完整性
Awareness Training	意识培训
Basis of Design	设计基准
Best Practice JHA	JHA范例
Buddy System	两人同行制
Business Continuity Plan	业务连续性计划
Closed Footwear	满帮鞋子
Crisis Management	危机管理
Critical Control Point (CCP)	关键控制点
Emergency	应急、紧急、紧急事态、紧急状况、紧急事件等
Emergency Control Centre(ECC)	应急控制中心
Emergency Management	应急管理
Emergency Response Centre(ERC)	应急响应中心
Emergency Response Plan(ERP)	应急响应计划

Emergency Situation	紧急事态
Evidence-Based Medicine (EBM)	循证医学
Exposure	暴露、照射、曝光、暴晒
Fitness to Work	人员适岗性
Food Handler	食品加工人员
Food Production	食品加工
Food Waste	厨余垃圾
Foodborne Illness	食源性疾病
General Cleanliness.	整体清洁情况
Global System for Moblie Communications (GSM)	全球移动通信系统
Hazard Analysis and Critical Control Point (HACCP)	危害分析与关键控制点
Hazard Identification (HAZID)	危害辨识
Hazard and Operability (HAZOP)	危险与可操作性
Health Risk Assessment (HRA)	健康风险评估
Health Surveillance	健康监护
Health Surveillance Programme	健康监护方案
Health Surveillance System	健康监护系统
Health, Safety and Environment Management System (HSE-MS)	健康安全环境管理体系
Health, Safety and Environment (HSE)	健康安全与环境
Hearing Conservation Programme (HCP)	听力保护计划
Hearing Protector	听力保护用品

Heat Illness Protocol	热病治疗规范
Heat Stress Management Programme (HSMP)	热应激管理方案
HSE Briefings	HSE择要说明(会)
HSE Critical Equipment and Systems (HSECES)	HSE关键设备和系统
In Vehicle Monitoring System(IVMS)	车载监控系统
Incident Command System (ICS)	突发事件应急指挥系统
Incident Management Team (IMT)	事件管理小组
International Association of Drilling Contractors(IADC)	国际钻井承包商协会
International Association of Oil & Gas Producers(IOGP)	国际油气生产商协会
International Good Practice	国际良好实践
Job Hazard Analysis (JHA)	工作危害分析
Journey Management Plan (JMP)	旅程管理计划
Lagging Indicator	跟随性指标
Layer of Protection Analysis (LOPA)	保护层分析
Leadership Site Visit (LSV)	现场安全督导
Leadership Site Visit Program(LSVP)	管理人员现场安全督导工作，或管理人员现场安全督导方案
Leading Indicator	前瞻性指标
Lone Worker	独自作业的人员
Lone Working	独自作业
Major Accident Hazard (MAH)	重大事故危害

Management of Change (MoC)	变更管理
Manual of Permitted Operations (MOPO)	许可操作手册
Medevac	医疗转运
Medical Emergency Response Plan (MERP)	医疗应急响应计划
Medical Evacuation	医疗转运
Method Statement	作业方法说明
Microbiological Pathogen	病原微生物/致病微生物
Night Working	夜间作业
Noise-Induced Deafness (NID)	噪声性耳聋
Occupational Safety and Health Administration (OSHA)	(美国)职业安全与卫生管理局
Override	旁路
Permit Control Facility	作业许可控制设施
Permit to Work (PTW)	作业许可
Personal Protective Equipment (PPE)	个人防护装备
Pest Control Programme	害虫控制方案
Pre-job Meeting	作业前会议
Pre-job Safety Talk	作业前安全交谈(谈话、喊话)
Process Isolation	工艺隔离
PTW System	作业许可系统
Quantitative Risk Assessment (QRA)	定量风险评估,定量风险分析
Refresher Training	更新培训
Safe Methods of Work	安全的工作方法

English	中文
Safe Operating Envelope	安全操作范围
Safety Case	安全例证
Safety Critical Personnel	安全关键人员
Safety Data Sheet for Chemical Products(SDS)	化学品安全技术说明书
Safety Integrity Level (SIL)	安全完整性等级
Self Contained Breathing Apparatus (SCBA)	正压呼吸器
Simultaneous Operations(SIMOPS)	同时作业,同步作业,交叉作业
Site Emergency Response Plan(SERP)	现场应急响应计划
Standard Operating Procedures(SOP)	标准作业程序
Subject Matter Expert (SME)	(某)领域专家或专业人员
(Supplementary) Certificate	专项作业单
Toolbox Talks (TBT)	工具箱会议
UK Health and Safety Executive(UK HSE)	英国健康安全执行局
Wear and Tear	磨损
Work Alone	独自作业
Work Instructions (WI)	作业指导书
Workplace	工作场所
World Health Organization (WHO)	世界卫生组织